INTERNATIONAL SYMPOSIUM ON ASTROPHYSICS RESEARCH AND ON THE DIALOGUE BETWEEN SCIENCE AND RELIGION

July 8-12, 2002
Vatican Observatory, Castel Gandolfo, Italy

EDITED BY CHRIS IMPEY AND CATHERINE PETRY

Hosted by the Vatican Observatory and
Sponsored by the Templeton Foundation

Cover Art

Foreground: Michelangelo Buonarroti, “Creation of the Sun and Moon, and Planets”, 1508-12. In this portion of Sistine Chapel Ceiling, Michelangelo depicts a part of the biblical story of the Creation. The panel shows God the Creator hovering above the Earth, robes and hair flying behind him as he moves in the void and creates the Sun and the Moon. The fresco is a remarkable example of Michelangelo's understanding of the human form and its movement (Courtesy: Cappella Sistina, Città del Vaticano).

Background: In this view of the Tarantula Nebula, newly formed stars are interspersed with curtains and filaments of hot gas. The Tarantula Nebula is located in the Large Magellanic Cloud, one of our nearest galactic neighbors at a distance of 170,000 light years. The red and green colors are caused by excited atoms of hydrogen and oxygen, respectively. The image was made with the WFPC2 camera on the Hubble Space Telescope (Courtesy: NASA and the Hubble Heritage Team, STScI/AURA; Y.-H. Chu, University of Illinois; and E. Grebel, University of Washington).

Cover Design Consultant: Carmen Ortiz Henley.

Table of Contents

Part 1. Keynote Papers

Part 2. Contributed Papers

Section A. Stars and Stellar Remnants

Section B. Galaxies and Cosmology

Section C. Religion and Smörgasbord

Foreword

An exciting and unusual meeting took place on the shores of Lake Albano, near Rome, in July of 2002. The first aspect was a research symposium for alumni of the Vatican Observatory Summer Schools (VOSS). Every two years, about two dozen young graduate students from all over the world come to the Vatican Observatory for a month of intense study and debate, and for an exhilirating immersion in Italian culture. Over the course of 9 schools since 1986, 225 students from 52 nations have received a career boost from VOSS. Sixty percent of the students are from developing countries, and 43% are women. Each school has a broad astrophysical theme and the young scholars are brought up to the research frontier of their subject by a select international faculty.

The first gathering of VOSS alumni took place in 1999. In 2002, 55 scholars from 22 countries presented their research. The symposium spanned the universe from meteorites to cosmology, and it was inspiring to see the career progression of successive generations of VOSS alumni. The second aspect of the meeting gave it a unique flavor – eight keynote speakers gave talks centered on the interface between science and religion. Ample time was allowed for discussion. Science and religion do not often rub shoulders, but when they do the results can be intellectually invigorating. The keynote presentations form the first part of this volume.

The meeting was successful thanks to the efforts of many people. Great thanks go to Rita Callegari, Elizabeth Maggio, and Melissa Pesce-Rollins for their outstanding work as secretaries of the conference. Many of the Vatican Observatory staff, especially Giuseppe Koch, S. J., Piero Piermattei, Luigi Lori, and Francesco Rossi, provided assistance, and staff at Mondo Migliore helped the conference sessions run smoothly. Funds for travel and organization came from the Vatican Observatory, and we are grateful to the Templeton Foundation for helping to make this event and the two subsequent publications possible. We are confident that this will not be the last time VOSS scholars gather to help dissolve the boundaries between disciplines and cultures.

George Coyne, S. J.	Host / Organizer
Chris Impey	Organizer / Editor
Catherine Petry	Editor

Conference Participants

A

Eun-Joo Ahn, Dept. of Astronomy & Astrophysics, University of Chicago, 5640 South Ellis Avenue, Chicago, IL 60637, USA (sein@oddjob.uchicago.edu)

Ayvur Akalin Peletier, 39 Woodland Grove, Beeston, Nottingham NG9 5BP, U.K. (ayvur_ peletier@hotmail.com)

Vladimir A. Avila Reese, Instituto de Astronomia-UNAM, Apartado Postal 70-264, 04510 Mexico D.F., Mexico (avila@astroscu.unam.mx)

B

Yury Bialetsky, Astronomical Observatory, Shevchenko Park, Odessa, 270014, Ukraine (yuri@deneb.odessa.ua)

Peter Biermann, Max Planck Institut, Auf dem Huegel 69, D-53121 Bonn, Germany (pl65bie@mpifr-bonn.mpg.de)

Ana Biscaya-Holzbach, 189 Webster St. #3, East Boston, MA 02128, USA (anaholzbach@yahoo.com)

Richard Boyle, Vatican Observatory Research Group, Steward Observatory, University of Arizona, Tucson, AZ 85721, USA (boyle@as.arizona.edu)

C

Rita Callegari Barlow, 206 Wingate Place, SW, Leesburg, VA 20175, USA (barlowd@mediasoft.net)

William Carroll, Tutorial Fellow in Theology, University of Oxford, UK (william.carroll@blackfriars.oxford.ac.uk)

Juan Casanovas, Specola Vaticana, V-00120, Vatican City, Rome, Italy (jcc@specola.va)

Miguel Chavez, INAOE, Departamento de Astrofisica, Apdo. Postal 216 y 51, 7200 Puebla, Pue, Mexico (mchavez@inaoep.mx)

Fernando Comeron Tejero, ESQ, Karl Schwarzschild Strasse 2, 85748 Garsching bei Muenchen, Germany (fcomeron@eso.org)

Guy Consolmagno, Specola Vaticana, V-00120, Vatican City, Rome, Italy (gjc@specola.va)

Sofia Cora, Observatorio Astronomico, Paseo del Bosque, 1900 La Plata, Argentina (sacora@fcaglp.fcaglp.unlp.edu.ar)

Christopher Corbally, Vatican Observatory Research Group, Steward Observatory, University of Arizona, Tucson, AZ 85721, USA (corbally@as.arizona.edu)

George Coyne, Specola Vaticana, V-00120, Vatican City, Rome, Italy (gcoyne@specola.va)

D

Francesca D'Antona, Oss. Astron. di Roma, Via dell'Osservatorio 1, I-00040 Monte Porzio (Rome), Italy (dantona@coma.mporzio.astro.it)

M. Devasagayam, S.J., Jesuit Residence, St. Joseph's College, Trichirappalli 620002, Tamil Nadu, India (devasagayam@hotmail.com)

Danuta Panek Dobrzycka, ESO, Karl Schwarzschild Strasse 2, 85748 Garsching bei Muenchen, Germany (danuta@cfa.harvard.edu)

Adam Dobrzycki, ESO, Karl Schwarzschild Strasse 2, 85748 Garsching bei Muenchen, Germany (adam@head-cfa.cfa.harvard.edu)

E

Philippe Eenens, Dept. de Astronomia, Universidad de Guanajuato, Apdo. Postal 144, C.P. 36000, Guanjuato, Gto., Mexico (eenens@avantel.net)

F

Diana Foss, 1571A Lincoln Ave., San José, CA 95125, USA (diana@mooseyard.com)

Jose Funes, Vatican Observatory Research Group, Steward Observatory, University of Arizona, Tucson, AZ 85721, USA (jfunes@as.arizona.edu)

G

Cristina Galea, Str. Straduintei Nr. 10, B1. R20, Sc. A, Et. 7, Ap. 31, Sect. 4, Bucharest, Romania, cod 75586 (cgalea@mpifr- bonn.mpg.de)

Javier A. García, Sector Panamericano, Calle 72 No. 52-88, Maracaibo 4005, Estado Zulia, Venezuela (jgarcia@cida.ve)

Robert Garrison, David Dunlap Observatory, University of Toronto, Box 360 Richmond Hill, Ontario L4C 4Y6, Canada (garrison@astro.utoronto.ca)

Owen Gingerich, Research Professor of Astronomy and History of Science, Harvard University, Cambridge, MA 02138, USA (ginger@cfa.harvard.edu)

Nadya Gorlova, Steward Observatory, University of Arizona, Tucson, AZ 85721, USA (ngorlova@as.arizona.edu)

I

Dragana Ilic, Slobodana Jovanovica 12/7, 11000 Belgrade, Yugoslavia (ida7@EUnet.yu)

Chris Impey, Steward Observatory, University of Arizona, Tucson, AZ 85721, USA (cimpey@as.arizona.edu)

K

Maria Marina Kaufman Bernardó, Laprida 1535 3 "G", (1425) Capital Federal, Argentina (mar@vianw.net.ar)

Giuseppe Koch, Specola Vaticana, V-00120, Vatican City, Rome, Italy (gkoch@specola.va)

L

Aigen Li, Lunar and Planetary Laboratory, University of Arizona, Tucson, AZ 85721 (agli@lpl.arizona.edu)

Omar López Cruz, Carnegie Observatories, 813 Santa Barbara Street, Pasadena, CA 91101 (omar@ociw.edu)

M

Witold Maciejewski, Astronomical Observatory, Jagiellonian University, ul. Orla 171, Fort Skala, PL 30-244 Krakow, Poland (witold@oa.uj.edu.pl)

Sabino Maffeo, Specola Vaticana, V-00120, Vatican City, Rome, Italy (smaffeo@specola.va)

Elizabeth Maggio, 6305 Monero Drive, Rancho Palos Verdes, CA 90275, USA (lizmaggio@compuserve.com)

Zaida Marin, CIDA, Apdo. Postal 264, Merida 5101-A, Venezuela (marin@cida.ve)

Ernan T. McMullin, Box 1066, University of Notre Dame, Notre Dame, Indiana 46556, USA (Ernan.McMullin.1@nd.edu)

Dante Minniti, Departamento de Astronomia, PUC, Avda. Vicuna Mackena 4860, Casilla 104, Santiago 22, Chile (dante@astro.puc.cl)

Felicitas Mokler, Max-Planck-Institut fuer Extraterrestrische Physik, Postfach 1312, D-85741 Garching, Germany (fmokler@mpe.mpg.de)

Christopher R. Mullis, ESO, Karl-Schwarzschild-Strasse 2, D-85748 Garching bei Muenchen, Germany (cmullis@eso.org)

Nancey Claire Murphy, Professor of Christian Philosophy, Fuller Theological Seminary, Pasadena, CA 91182, USA (nmurphy@fuller.edu)

P

Melissa Pesce-Rollins, Via Fucini 19, Campiglia Mma., 57021 Livorno, Italy (campiglia44@hotmail.com)

Kristof Petrovay, Eötvös University, Budapest, P. O. Box 32, H-1518 Hungary (k.petrovay@astro.elte.hu)

R

Lynn J. Rothschild, Ecosystem Sci. & Tech. Branch, Mail Stop 239-20, NASA Ames, Moffett Field, CA 94035, USA (Lynn.J.Rothschild@nasa.gov)

S

Sarah N. Sägesser, Lincoln College, Turl Street, Oxford, OX1 3DR, UK (sagesser@thphys.ox.ac.uk)

Lousine A. Sargsysan, (Sar11@yahoo.com)

Alejandro Soto, 385 Stonebridge Trail, Roswell, GA 30075, USA (Alejandro.M.Soto.97@Alum.Dartmouth.ORG)

John T. Stocke, Department APS, Campus Box 391, University of Colorado, Boulder, CO 80309, USA (stocke@casa.colorado.edu)

T

Trinh Xuan Thuan, Department of Astronomy, University of Virginia, Charlottesville, VA 22903, USA (txt@bluecompact.astro.virginia.edu)

Juraj Toth, Faculty of Mathematics Physics and Informatics, Comenius University, Bratislava, Slovak Republic (Toth@center.fmph.uniba.sk)

V

Leonardo Vanzi, European Southern Observatory, Alonso de Cordona 3107, Vitacura, Santiago, Chile (lvanzi@eso.org)

Simon Vidrih, Department of Physics, University of Ljubljana, Jadranska 19, Ljubljana, Slovenia (simon.vidrih@fmf.uni-lj.si)

W

Yiping Wang, Purple Mountain Observatory, Academia Sinica, Nanjing 210008, China (ypwang@pmo.ac.cn)

Andrew Whitman, Vatican Observatory Research Group, Steward Observatory, University of Arizona, Tucson, AZ 85721, USA (whitman@as.arizona.edu)

Part 1. Keynote Papers

International Symposium on Astrophysics Research and on the Dialogue between Science and Religion, Vatican Observatory, 2003
C. D. Impey & C. E. Petry, editors

Thomas Aquinas, Creation and Big Bang Cosmology

William E. Carroll
University of Oxford

Introduction

Thomas Aquinas and Big Bang Cosmology may seem to be an odd juxtaposition. What relation could there be between a theologian and philosopher of the thirteenth century and the widely accepted cosmological theory of our own century? What I hope to show is that Thomas Aquinas has a great deal to say as to how we ought to interpret claims in contemporary science, especially when those claims involve, or seem to involve, the origin of the universe.

Recent studies in particle physics and astronomy have produced dazzling speculations about the early history of the universe. Cosmologists now routinely entertain elaborate scenarios which propose to describe what the universe was like when it was the size of a grapefruit, a mere 10^{-35} seconds after the Big Bang. The description of the emergence of four fundamental forces and twelve discrete subatomic particles is almost a common place in modern physics. There is little doubt among scientists that we live in the midst of a great expansion which began 15 billion years ago – give or take a few billion.

The Nature of the Big Bang

Stephen Hawking has observed that, as a result of contemporary cosmology, the question of the beginning of the universe has entered "the realm of science". Hawking has argued that we can have no scientific theory of nature unless the theory accounts for the beginning of the universe.

The only way to have a scientific theory is if the laws of physics hold everywhere, including at the beginning of the universe. One can regard this as a triumph of the principles of democracy: why should the beginning of the universe be exempt from the laws that apply to other points? If all points are equal, one can't allow some to be more equal than others.[1]

This confidence that cosmology now can address the beginning of the universe – a confidence shared by many cosmologists – has led to all sorts of speculations about the initial state of the universe. For many scientists, philosophers, and theologians such speculations in cosmology speak directly to long-established beliefs about creation. In January 2001, the science editor of *The New York Times* wrote that high-speed particle accelerators may help scientists to work out "a mechanistic, gears-and-levers theory of the Genesis moment itself – the hows if not the whys of creation *ex nihilo*".

Most physicists refer to the Big Bang as a "singularity", that is, an ultimate boundary or edge, a "state of infinite density" where space-time has ceased. Thus, it represents an outer limit of what we can know about the universe. If all physical theories are formulated in the context of space and time, it would not be possible to speculate, at least in the natural sciences, about conditions before or beyond these categories. Nevertheless, during the last twenty years, precisely such speculation has intrigued several cosmologists. Some of them now offer theories that propose to account for the Big Bang itself as a fluctuation of a primal vacuum. Just as sub-atomic particles are thought to emerge spontaneously in vacuums in laboratories, so the whole universe may be the result of a similar process.

Alexander Vilenkin of Tufts University has developed a variation of an inflationary model of the expanding universe that accounts for the birth of the universe "by quantum tunneling from nothing". "Nothing", for Vilenkin, is a "state with no classical space-time… the realm of unrestrained quantum gravity; it is a rather bizarre state in which all our basic notions of space, time, energy, entropy, and the like lose their meaning".[2] Describing these speculations in a recent book, *The Inflationary Universe,* Alan Guth appropriates traditional theological terminology in a chapter called: "A Universe *ex nihilo*".

Are we on the verge of a scientific explanation of the very origin of the universe? The contention of several proponents of the new theories is that the laws of physics are sufficient to account for the origin and existence of the universe. If this is true, then, in a sense, we live in a universe that needs no explanation beyond itself, a universe which has sprung into existence spontaneously from a cosmic nothingness.

The philosopher Quentin Smith writes: "there is sufficient evidence to warrant the conclusion that the universe… began to exist without being caused to do so". The title of his essay is "The Uncaused Beginning of the Universe", and his conclusion is revealing: "…the fact of the matter is that the most reasonable belief is that we came from nothing, by nothing and for nothing". He writes that if Big Bang cosmology is true "…our universe exists without cause or without explanation… [This world] exists nonnecessarily, improbably, and causelessly. It exists for *absolutely no reason at all*".[3]

There is another major trend in the application of quantum mechanics to cosmology – different from the inflationary universe and the quantum tunneling from nothing described by Vilenkin – but no less significant in the claims it makes, or the claims that are made for it, concerning the answers to ultimate questions about the universe. This is the view made famous by Stephen Hawking, who, in *A Brief History of Time,* has taken a different route: challenging the very notion of an absolute beginning to the universe. He denies the intelligibility of an event that would, in principle, be beyond the realm of scientific investigation. There is no singularity, no initial boundary at all; the universe has no beginning!

Any appeal to an initial singularity is, for Hawking, an admission of defeat: "If the laws of physics could break down at the beginning of the universe, why couldn't they break down anywhere?" To admit a singularity is to deny a universal predictability to physics, and, hence ultimately, to reject the competency of science to understand the universe. Hawking is not shy at all about drawing theological conclusions from his cosmological speculations. The universe is "completely self-contained and not affected by anything outside itself". In Hawking's cosmology, the universe "would neither be created nor destroyed. It would just *be*".[4] If the universe had no beginning, there is nothing whatsoever for God to do – except, perhaps, to choose the laws of physics. Physics, were it to embrace a unified theory, will allow us to know "the mind of God". Here are Hawking's words:

> So long as the universe had a beginning, we could suppose it had a creator. But if the universe is really completely selfcontained, having no boundary or edge, it would have neither beginning nor end: it would simply be. What place, then, for a creator?[5]

In his book *The Universe in a Nutshell,* Hawking observes that the cosmological theory he sets forth concerning the very early history of the universe means that the universe is "entirely self-contained", it "doesn't need anything outside to wind up the clockwork and set it going. Instead, everything in the universe [is]… determined by the laws of science and by the rolls of the dice within the universe. This [conception] may sound presumptuous, but it is what I and many other scientists believe".[6]

Andrei Linde, a cosmologist at Stanford University, speculating on what he admits is a bizarre question – what happened before the Big Bang – has developed a theory of "eternal inflation", according to which what we know as the Big Bang is only one of many in a chain of big bangs by which "the universe endlessly reproduces and reinvents itself". According to Linde, "our universe" began as a bubble that ballooned out of the space-time of a pre-existing universe. He thinks that it makes little sense to search for some "original bubble".[7]

Another theoretical physicist, Lee Smolin, imagines a whole chain of universes that develop according to the theory of "cosmological natural selection", so that "our universe forms part of an endless chain of self-reproducing universes whose physical laws evolve as they are passed along". For Smolin, "the laws of physics in this universe (or universes) are less like commandments from God and more like the zoning regulations promulgated by some fractious city council, ever susceptible to amendment and compromise". He thinks that the universe is like a city, "an endless negotiation, an endless construction of the new out of the old… No one made the city. There is no city-maker, as there is no clockmaker. If a city can make itself without a maker, why cannot the same be true of the universe?" Each black hole, just like the black hole in which the Big Bang occurred, begets a new universe which expands, evolves, and eventually creates new black holes which spawn new universes: "…over many cycles a kind of Darwinian pressure would encourage the formation of universes whose physics favored black holes, since universes that did not make black holes would have no progeny".[8]

The Role of the Creator

It is interesting that some Christians rushed to embrace traditional Big Bang cosmology because they saw it as scientific confirmation of the Genesis story of creation. Accordingly, we may understand the particular attraction of some thinkers to current variations in cosmology that purports to account for the initial singularity in terms of quantum tunneling, or to deny the existence of an initial boundary to the universe, or to speak of an eternal cycle of big bangs. In each case, so it might seem, the role of a creator is superfluous. But, as we shall see, to use cosmological theories either to affirm creation or to deny it is an example of misunderstandings of both cosmology and creation.

The universe described by Hawking, and others – the fruit so it seems of contemporary cosmology – is a self-contained universe, exhaustively understood in terms of the laws of physics. In such a universe there would seem to be little if any need for the God of Jewish, Christian, or Muslim revelation. The traditional doctrine of creation seems obsolete in the face of the recent advances of science.[9] For some people, the notion of a Creator represents an intellectual artifact from a less enlightened age.

Too often, contemporary discussions about the theological and philosophical implications of Big Bang cosmology, as that cosmology has been refined, suffer from an ignorance of the history of science, and, with respect to the theories which claim to involve the origin of the universe, these recent discussions reveal an ignorance of the sophisticated analyses of the natural sciences and creation which took place in the Middle Ages. The reception of Aristotelian science in Muslim, Jewish, and Christian intellectual circles in the Middle Ages provided the occasion for a wideranging discussion of the relationship between theology and the natural sciences. Thomas Aquinas' understanding of creation – and, in particular, the distinctions he draws among theology, metaphysics, and the natural sciences – can continue to serve as an anchor of intelligibility in a contemporary sea of speculative theories.[10]

Something from Nothing

It seemed to many of Thomas' contemporaries that there was a fundamental incompatibility between the claim of ancient physics that something cannot come from absolutely nothing and the affirmation of Christian faith that God did produce everything from nothing. Furthermore, for the ancients, since something must come from something, there must always be something, i.e., the universe must be eternal. Despite the claims of some contemporary theorists that, properly speaking, we can get something from nothing, those theories of the Big Bang that employ insights from particle physics concerning vacuum fluctuations[11] are really consistent with the ancient principle that you cannot get something from nothing.

A persistent confusion between cosmological and philosophical conceptions of "nothing" is evident in these discussions.[12] The "vacuum" of modern particle physics, whose "fluctuation" supposedly brings our universe into existence, is not absolutely nothing. It may be no thing like our present universe, but it is still something. How else could "it" fluctuate? So the "nothing" of contemporary cosmological theories turns out to be really something. Yet, the notion of "nothing", central to the theological and metaphysical conception of creation out of nothing, is radically different from the various notions of "nothing" employed in the contemporary cosmological discourse.[13] To speak of "creation out of nothing" means that one is denying that any matter at all is changed or transformed into something else. The expression "out of nothing" or "from nothing" is, at its root, a denial of any material cause whatsoever in the act of creation. Categories of explanation concerning creation are proper to metaphysics and theology, not the natural sciences. Examples of this confusion can also be found in some theologians who have gone to considerable lengths to find a concordance between "quantum tunneling from nothing" and the opening of Genesis. They find the eruption of being from a primal vacuum to be remarkably analogous to the second line of Genesis according to which the Spirit of God hovers over the abysmal deep and brings forth our world. "Such a God is not an *entity* who throws the switch of creation, but the amazing *fact* that something actually seethes into being from nothing".[14]

For many mediaeval thinkers, the eternal universe of ancient Greek science seemed to be incompatible with a universe created out of nothing. An eternal universe is, so it seemed, a necessary universe, a universe that is not the result of the free creative act of God. At least so some Christians thought, and they urged that ancient science, especially in the person of Aristotle, its leading proponent, be banned, since it contradicted the truths of revelation.[15] Muslim and Jewish scholars had already wrestled with the heritage of Greek science as they sought to understand what it meant to believe in God as Creator. But it was Thomas Aquinas who succeeded in forging a robust understanding of creation out of nothing that honored both the requirements of biblical revelation and a scientific account of nature. As I have suggested, I think that Thomas' contribution to the mediaeval discussion of creation and science speaks directly to discourse on cosmology and creation in our own day.

Creation and Change

The key to Thomas Aquinas' analysis is the distinction he draws between the act of creation and change, or as he often said: *creatio non est mutatio* (creation is not a change). The natural sciences, whether Aristotelian or contemporary, have as their subject the world of changing things: from subatomic particles to acorns to galaxies. Whenever there is a change there must be something that changes. The ancients were right: from nothing, nothing comes; that is, if the verb "to come" means a change. All change requires some underlying material reality.

To create, on the other hand, is to cause the *whole* reality of whatever exists. To cause completely something to exist is not to produce a change in something; to create, thus, is not to work on or with some already existing material. If there were a prior something that was used in the act of producing a new thing then the agent doing the producing would not be the *complete* cause of the new thing. But such a complete causing is precisely what the act of creation is. Thus, to create is to give existence, and all things depend upon God for the fact that they are. God does not take nothing and make something. Rather, any thing left entirely to itself, separated from the cause of its existence, would be absolutely nothing. Creation is not exclusively, nor even primarily, some distant event; God's creative act is the continual, complete causing of the existence of whatever is.[16]

Thomas remarks that the relation of a house to its builder is very different from the relation of a creature to the Creator. Once the coming-to-be of the house is complete, the house ceases to have any relation of dependence upon its builder; the builder could die, and the house would continue to stand. But the case is quite otherwise with the creature *qua* creature. The Creator's causality must be continual, and of the same kind, all throughout the creature's existence. All things would fall into non-being, Thomas says, unless God's omnipotence supported them.

Thomas is particularly insightful in distinguishing between the origin of the universe and the beginning of the universe. Beginning refers to a temporal event, and an absolute beginning of the universe would be an event that is coincident with the beginning of time. Creation is an account of the origin, or source of existence, of the universe, and, as such, Thomas thinks that creation can be demonstrated in the science of metaphysics. In his *Writings on the Sentences of Peter Lombard,* completed in Paris in the 1250s, Thomas claims: "Not only does faith hold that there is creation, but reason also demonstrates it".[17] The development by Thomas of an understanding of creation *ex nihilo,* and, in particular, his understanding of the possibility of an eternal, created universe, offers one of the best examples of the relationship between faith and reason. In fact, his magisterial treatment of the doctrine of creation is one of the enduring accomplishments of the thirteenth century.

For Thomas there are two senses of creation out of nothing, one philosophical, the other theological. The philosophical sense simply means that God, with no material cause, makes all things to exist as entities that are really different from Him, yet completely dependent upon His causality. The theological sense of creation denies nothing of the philosophical sense and adds to it the notion that the created universe is temporally finite. I do not intend here to provide the metaphysical argument by which Thomas moves from the distinction between the essences of things and their existence, to the conclusion that there *must be* an uncaused cause of the existence of all that is. I only want to point out that, for Thomas, reason alone can arrive at an understanding of many of the essential features of the doctrine of creation.

The Beginning and the End

Thomas observes that "the causality of the Creator... extends to everything that is in the thing. And, therefore, creation is said to be *out of nothing,* because nothing uncreated pre-exists creation".[18] The Creator is prior to what is created, but the priority is not *fundamentally* temporal. Each creature has its origin in the Creator and is wholly dependent upon the Creator for its existence; the dependence is metaphysical not temporal. To be created out of nothing does not mean that the creature is *first* nothing and *then* something.

Thomas Aquinas saw no contradiction in the notion of an eternal created universe. For, even if the universe had no temporal beginning, it still would depend upon God for its very being. The radical dependence on God as cause of being is what creation means.

Thomas would note that to argue that the universe has no beginning (either because it is eternal as the ancients thought, or because the very notion of temporality is a subsidiary concept as Hawking thinks) does not challenge the fundamental metaphysical truth that the universe has an origin, i.e., that the universe is created. Whether there is "eternal inflation", as Andrei Linde thinks, or perhaps an endless series of universes within the universe, all such universes would still require God's creative act in order to exist.

There is no necessary conflict between the doctrine of creation and any physical theory. Theories in the natural sciences account for change. Whether the changes described are biological or cosmological, unending or temporally finite, they remain processes. Creation accounts for the existence of things, not for changes in things. As Thomas says: "Over and above the mode of becoming by which something comes to be through change or motion, there must be a mode of becoming or origin of things without any mutation or motion through the influx of being".[19]

There were some thinkers in the Middle Ages, in each of the three great religious traditions, who thought that science could demonstrate that the universe had a temporal beginning. Such confidence in our ability to know that the universe is temporally finite can be seen as well in the arguments of several Big Bang cosmologists. Thomas, however, following the lead of Maimonides, argued that, in principle, science can not conclude that the universe has a temporal beginning. Although, as we have seen, Thomas did think that reason can demonstrate that the universe has an origin, that is, that the universe is radically dependent upon a cause for its existence, he thought that it was an error to think that, on the basis of how we understand the universe in its current state, we can extrapolate or reason to an initial state or temporal beginning of the universe. Thomas observes:

> ...that the world had a beginning... is an object of faith, but not a demonstration or science. And we do well to keep this in mind; otherwise, if we presumptuously undertake to demonstrate what is of faith, we may introduce arguments that are not strictly con-

> clusive; and this would furnish infidels with an occasion for scoffing, as they would think that we assent to truths of faith on such grounds.[20]

Contrary to Hawking's observation that I quoted earlier in this article, there are different senses of beginning: an absolute temporal beginning of the universe would be quite unlike any beginning that occurs in the universe.

Consulting the Bible

Thomas did *believe* that the Bible revealed the universe is not eternal; the Fourth Lateran Council in 1215 had formally defined this doctrine. Aristotle, Thomas thought, was wrong to think that the universe was eternal. But he argued that, on the basis of reason alone, one could not know whether the universe is eternal. To affirm, on the basis of faith, that the universe has a temporal beginning, involves no contradiction with what the natural sciences can legitimately proclaim. Since the natural sciences cannot know whether the universe has a temporal beginning, a revelation in faith on this subject completes and perfects what reason knows. Thomas, however, would find faulty those arguments that sought to show that the world is created on the basis of the "singularity" of traditional Big Bang cosmology.

I should like to make an additional, brief point concerning Thomas' understanding of the Bible on the question of creation. Some defenders as well as critics of contemporary science, especially evolution, think that belief in the Genesis account of creation is incompatible with science. Thomas, however, did not think that the Book of Genesis presented any difficulties for the natural sciences, for the Bible is not a textbook in the sciences. What is essential to Christian faith, according to Thomas, is the "fact of creation", not the manner or mode of the formation of the world. Thomas notes:

> There are some things that are by their very nature the substance of faith, as to say of God that He is three and one... about which it is forbidden to think otherwise... There are other things that relate to the faith only incidentally... and, with respect to these, Christian authors have different opinions, interpreting the Sacred Scripture in various ways. Thus with respect to the origin of the world, there is one point that is of the substance of faith, *viz.*, to know that it began by creation... But the manner and the order according to which creation took place concerns the faith only incidentally.[21]

Thomas' firm adherence to the truth of Scripture without falling into the trap of literalistic readings of the text offers valuable correction for exegesis of the Bible which concludes that one must choose between the literal interpretation of the Bible and modern science. For Thomas, the literal meaning of the Bible is what God, its ultimate author, intends the words to mean. The literal sense of the text includes metaphors, similes, and other figures of speech useful to accommodate the truth of the Bible to the understanding of its readers. For example, when one reads in the Bible that God stretches out His hand, one ought not to think that God has a hand. The literal meaning of such passages concerns God's power, not His anatomy.

It is also important to recognize the distinction between creation, understood as God's causing the universe to be, and the account of the "six days of creation" set forth in the Bible. Beginning with the Church Fathers there have been many attempts to discover a concordance between the description of the formation of the world found in the early chapters of Genesis and scientific knowledge. Such commentaries on Genesis are fundamentally different from a theological or a philosophical analysis of creation, which concerns the dependence of all that is on God.

Existence and Creation

The Big Bang described by modern cosmologists is not creation. The natural sciences cannot themselves provide an ultimate account for the existence of all things. It does not follow, however, that reason remains silent about the origin of the universe. Reason embraces more than the categories of the natural sciences. As we have seen, although Thomas did not think that reason alone can conclude that the universe has a temporal beginning, he did think that reason alone can demonstrate that the universe is created. Such a demonstration occurs in metaphysics. Just as metaphysics ought not to deny the truths about the world discovered in the natural sciences, so too the natural sciences ought not to reject the truths discovered in metaphysics. We must remember that it is one thing to attend to the processes that occur in nature; it is another to examine what it means for things to exist at all.

In *Three Roads to Quantum Gravity,* Lee Smolin provides an excellent example of the reduction of all explanations of the universe to the domain of the natural sciences:

> We humans are the species that makes things. So when we find something that appears to be beautifully and intricately structured, our almost instinctive response is to ask, "Who made that?" The most important lesson to be learned if we are to prepare ourselves to approach the universe scientifically is that this is not the right question to ask. It is true that the universe is as beautiful as it is intrinsically structured. But it cannot have been made by anything that exists outside of it, for by definition the universe is all there is, and there can be nothing outside it. And, by definition, neither can there have been anything before the universe that caused it, for if anything existed it must have been part of the universe. So the first principle of cosmology must be "There is noth-

> ing outside the universe". ...The first principle means that we take the universe to be, by definition, a closed system. It means that the explanation for anything in the universe can involve only other things that also exist in the universe.[22]

Although it is true that the cause of the existence of the universe is not a proper question for a cosmologist, it does not follow that the universe "cannot have been made by anything that exists outside of it". A universe that is the result of the fluctuation of a primal vacuum is not a self-creating universe. Nor is this primal vacuum the nothingness affirmed in creation out of nothing. When Lee Smolin, commenting on the philosophical and theological implications of current cosmological theories, writes "there never was a God, no pilot who made the world by imposing order on chaos and who remains outside, watching and proscribing",[23] he misunderstands both God and what it means to create. Contrary to the claim that the universe described by contemporary cosmology leaves nothing for a Creator to do, were a Creator not causing all that is, there would be nothing done!

Stephen Hawking is also wrong to conclude that there are implications for God as creator "if the universe is completely self-contained, with no singularities or boundaries, and completely described by a unified theory". One mistake which Hawking and others make in their denial of creation is the old error – which Thomas pointed out – of thinking that *ex nihilo* (out of nothing) *necessarily* means *post nihilum* (after nothing). Thus, by denying the latter (that creation occurs *after* nothing), they think that they also deny the former (creation *out of* nothing). Another mistake they make is to think that to create means to be an agent cause of change. Hawking denies that there is an initial change – his universe has no initial boundary, no beginning – thus, he thinks there is no active role for God to play. But since creation is not a change, Hawking's speculations do not really deny God's creative agency. Similarly, the "self-reproducing universes" of some cosmologists are not self-creating universes.

In August 2001, when Britain's Secretary for Trade and Industry, Patricia Hewitt, did something that would make her the envy of fellow politicians everywhere. At the University of Durham in England, she switched on a powerful new computer, popularly known as the "Cosmology Machine", and with the click of a mouse she was able to cause the computer to simulate a universe of her own on the screens of the University's Institute for Computational Cosmology. "Gosh", she said. "We must not let this go to our heads". Professor Carlos Frenk, Director of the Institute, said that the new computer had been "taught" the laws of physics and that its computational power would be used to simulate a number of virtual universes. *The Manchester Guardian* claimed that the expectation is that the new computer will disclose how the universe "inflated itself out of nothing in an instant". Despite the impressive calculatory capacity of the new computer, scientists may well be waiting quite some time for this disclosure!

To speak of the universe's *inflating itself* "out of nothing in an instant" captures much of the confusion in contemporary discussions about the implications of recent cosmology. But, as we have seen, the need to explain the existence of things does not disappear as a result of new explanations that propose to account for various changes (or even to deny them), regardless of how ancient or primordial these changes are. Thomas Aquinas would have no difficulty accepting Big Bang cosmology, even with its recent variations, while also affirming the doctrine of creation from nothing.

He would, of course, distinguish between advances in the natural sciences and the philosophical and theological reflections on these advances.

The variations in current cosmology to which I have referred are only theoretical speculations, and they are likely to change. To speculate, however, does not justify failures to make distinctions among the domains of the natural sciences, metaphysics, and theology. Nor does it justify fanciful philosophical and theological conclusions about a universe without cause. Thomas Aquinas did not have the advantage of the Hubble Space Telescope or the University of Durham's "Cosmology Machine", but in many ways he was able to see farther and more clearly than those who do.

NOTES

[1] Stephen W. Hawking and Roger Penrose, *The Nature of Space and Time* (Princeton: Princeton University Press, 1996), p. 71.

[2] "Birth of Inflationary Universes", in *Physical Review D,* 27:12 (1983), p. 2851.

[3] William Lane Craig and Quentin Smith, *Theism, Atheism and Big Bang Cosmology* (Oxford: Oxford University Press, 1993), p. 109.

[4] Stephen W. Hawking, *A Brief History of Time: From the Big Bang to Black Holes* (New York: Bantam Books, 1988), p. 136.

[5] *Ibid.,* p. 141.

[6] Stephen W. Hawking, *The Universe in a Nutshell* (New York: Bantam Books, 2001), p. 85.

[7] Quoted in Dennis Overbye, "Before the Big Bang, There Was... What?" *The New York Times,* 22 May 2001. Overbye offers an excellent *tour d'horizon* of current cosmological speculations: from quantum tunneling from nothing, to eternal inflation, to string theory and multiple universes, to Neil Turok's (Cambridge University) "ekpyrotic" universe [from "ekpyrosis", which denotes the fiery death and rebirth of the world in Stoic philosophy], to Linde's modification, called the "pyrotechnic universe".

[8] Lee Smolin quoted in an interview in Dennis Overbye, "The Cosmos According to Darwin", *The New York Times Magazine,* 13 July 1997, p. 26 and 27.

[9] For a discussion of these reactions, see William E. Carroll, "Big Bang Cosmology, Quantum Tunneling from Nothing, and Creation", *Laval théologique et philosophique,* 44, no. 1 (février 1988), pp. 59-75, at pp. 64-67, and William E. Carroll, "Thomas Aquinas and Big Bang Cosmology", *Sapientia* 53 (1998), pp. 73-95.

[10] For a discussion of Thomas Aquinas' understanding of creation, see Steven E. Baldner and William E. Carroll, *Aquinas on Creation* (Toronto: Pontifical Institute of Mediaeval Studies Press, 1997).

[11] According to John Gribbin, "The quantum vacuum is a seething froth of particles, constantly appearing and disappearing, and giving 'nothing at all' a rich quantum structure. The rapidly appearing and disappearing particles are known as virtual particles, and are said to be produced by quantum fluctuations of the vacuum". *In the Beginning: After COBE and Before the Big Bang* (Boston, Mass.: Little, Brown, and Company, 1993), pp. 246-7.

[12] Andre Linde remarked "at some moment" billions of years ago, "a tiny speck of primordial nothingness was somehow filled with intense energy with bizarre particles". One wonders at what greater size this "primordial speck of nothingness" would be "something" rather than "nothing!" *The New York Times,* 6 February 2001.

[13] This point is made especially clear in Michael Heller, "Cosmological Singularity and the Creation of the Universe", *Zygon,* Vol. 35, No. 3 (September 2000), pp. 665-685, at p. 670. Heller is also insightful in his analysis of the mathematical formalism involved in various notions of singularity and the problems concerning the relationship between such mathematical analysis and the physical universe.

[14] Paul Brockelman, *Cosmology and Creation: The Spiritual Significance of Contemporary Cosmology* (1999).

[15] Luca Bianchi, *L'errore di Aristotele: La polemica contro l'eternità del mondo nel XIII secolo* (Firenze: L. Olschki, 1984); Richard C. Dales, *Medieval Discussions of the Eternity of the World* (Leiden: Brill, 1990).

[16] Many theologians and philosophers find considerable significance in a distinction between an original act of creation and God's continuing causal agency. But, for Aquinas, there is really no difference between creation and what is called conservation; conservation is simply the continuation of creation. "It ought to be said that God does not produce things into being by one operation and conserve them in being by another. The being *[esse]* of permanent things is not divisible, except accidentally as it is subject to some motion; being, however, exists in an instant. Whence the operation of God does not differ according as it makes the beginning of being and as it makes the continuation of being". *De potentia Dei,* q. 5, a. 1, ad. 2. The reason given here for the fact that creation and conservation are the same is not that in God all things are one in His perfect simplicity, but that the effect of God's causality, the being of the creature, is the same effect all throughout the existence of the creature.

[17] We find Aquinas' claim that reason demonstrates creation in several places, two of which are of particular importance: *In II Sent. 1, 1,* ea. 1 and 2; *De potentia Dei,* 3, 5. In this latter text, Aquinas combines two separate arguments from Aristotle: the first is an argument from participation taken from *Metaphysics* 2.1 (993b23-27); the second, the argument from motion to a first unmoved mover, taken from books seven and eight of the *Physics. By* means of the second argument Aquinas proves the existence of a most perfect and true being; by means of the first argument he proves that all other things participate in the most perfect and true being.

[18] *In II Sent., 1,* 1, 2, resp.

[19] *De substantiis separatis,* c. 9, n. 49.

[20] *Summa theologiae I,* q. 46, a. 2.

[21] *In II Sent.,* dist 12, q. 3, a. 1.

[22] *Three Roads to Quantum Gravity* (New York: Basic Books, 2001), p. 17.

[23] Lee Smolin, *The Life of the Cosmos* (Oxford: Oxford University Press, 1998), p. 299.

International Symposium on Astrophysics Research and on the Dialogue between Science and Religion, Vatican Observatory, 2003
C. D. Impey & C. E. Petry, editors

The Sacred Cows of Religion and Science Meet

George V. Coyne, S. J.
Vatican Observatory

The Adventure of Looking Up

When we try to draw some implications for religious thought from our scientific experience, or when we proceed in the contrary direction to explore what science has to say to religious culture, we are setting out on an adventure, one in which no one area of human thought and experience can claim proprietorship and one in which much that is tentative, at times even tendentious, will come to light. All of us are aware that our cultivated ignorance and our tentative proposals may hopefully bring us closer to the truth in this dialogue, but only if we exercise a degree of intellectual discipline, perhaps even more than is typical of our respective professional areas of inquiry.

But this is not only an intellectual adventure. It is rather a human adventure in which the emotional and psychological part of us must also have its say. When we human beings think about ourselves in comparison with all else we experience in the universe we recognize that we have much in common and yet that we are different. More than any other creatures with whom we are familiar we sense that we are open to an immense and rich array of possibilities in the way we exist and in the way we lead our lives. This is, perhaps, best described by saying that we are symbolic creatures. Like symbols we are always leaning towards a reality beyond ourselves, towards the future and the unexplored. We are never fully content with what we are now. We are directed to the other and to the future.

Among the many symbolic, future-directed actions that express the nature of humans, one of the most meaningful and at the same time simple, is our gaze towards the sky in search of familiarity and understanding. That gesture of raising one's head and lifting one's eyes to gaze beyond immediate realities does not so much express a disorientation as it does an acknowledgment of insufficiency, of the need for something or someone out there, beyond oneself. Ancient mythologies, cosmologies, and cosmogonies bear witness to the immense power that drives humans in our continuous search for deeper understanding. Modern science bears witness to this persistent journey towards further understanding and the desire to accomplish something of significance.

There is little doubt that science has been a dominating influence in establishing the way we think today about the universe and ourselves. Nevertheless, we have become aware, in our attempt to unify our scientific knowledge with all that we have come to experience as human beings, that there is much in our experience that lies outside the domain of science and which today, more than ever in the past, throws open the doors to realities before which we sense that science itself is not totally competent, realities which require different approaches from those with which science itself is familiar. Science today is ever more human; it stimulates, provokes, and questions us in ways that drive us beyond science in the search for satisfaction, while at the same time scientific data furnish the stimuli. In this context the best science, to its great merit, neither pretends nor presumes to have the ultimate answers. It simply suggests and urges us on, well aware that not all is within its ken. Freedom to seek further understanding, and not a dogmatic possession of that which is partially understood, characterize the best scientists. Science, in fact, is a field where certainties lie always in the future; thus science is vital, dynamic, and very demanding of those who seek to discover the secrets of the universe and themselves.

In an attempt to establish a dialogue with religion we should read the data of science more from a human than a scientific perspective without any presupposition, however, that one precludes the other. In fact, it is altogether a very human endeavor to attempt to integrate rigorously scientific conclusions with those that are drawn from life's other experiences and from the very search for the meaning of life.[1] However, we should have no illusion, even those of us who are religious believers, that we possess the truth; quite the contrary, we are very aware that the truth is not to be possessed but to be contemplated. Contemplation is an end in itself and the object contemplated serves no other end than to bring the joy of contemplation to the observer. The truth itself requires that we neither exclude nor absolutize any possibility, if for no other reason than that our search be complete.

History, however, testifies to a much different view of both religion and science. And to a great extent that history lives on. Both religion and science have, to some extent, become "sacred cows".

A Misadventure: The Sacred Cows

The phrase "Sacred Cows" refers to a story in the Sacred Scriptures of the Judeo-Christian tradition about Moses' leadership over the Jewish people. Moses came down from the mountain and found the people dancing around and worshiping a statute of a calf made of gold. The story has several very important elements such as the following. As Moses prepared to go up the mountain the people asked him to find out what God's name was. We must appreciate what this meant to them. Scripture scholars have studied this in detail.[2] To the Jewish people, chosen by God at that time, name meant a very special thing. To them to possess the name of a person indicated a certain power to control that person. To possess the name of someone or something meant to bring that someone or something under your domination, within your ambience. So that is why they asked Moses to find out God's name. They would then know how to control God, how to pray to him so that he would answer them.

When, therefore, Moses talked to God, who spoke out of the burning bush, he asked God what his name was. God responded in a very cryptic way. He said to Moses, "You may tell the people I AM WHO AM". That is exactly the acronym YAHWEH. This is interpreted by Scripture scholars to have both a very negative and a very positive meaning. The very negative meaning is that "I Am Who I Am" signifies that I cannot give you my name. You cannot name me like you can Tree, or Fish, or Judy or Robert, or Galaxy or any other such object or person. I am not one of the many objects or many people whom you know. I cannot give you my name. But the phrase also has a very positive meaning. "I Am Who Am", means that I am so confident of myself, I know myself so well, that I know how much I love you. And even though you will turn against me, you are my chosen people. I will get angry, but I will never separate myself from you. I will always be your God. So there is both the positive and the negative meaning to God's response to Moses: I will always be faithful to you, but you cannot control me.

So Moses came down from the mountain and he found the people doing exactly what God had indicated to him they would do: they were bringing their god under their control. They could dance around their god, they could sing to their god, they could give their god a name by naming a statue, the Golden Calf. So they were making God in their own image and likeness, rather than what Scripture tells us: that we are made in the image and likeness of God.

The important thing here from the religious point of view is that the first motion in any religious relationship comes from God himself. God makes the first move. We do not reason our way to God, we do not earn our way to God, we do nothing to deserve that God gives himself to us and calls us into the ranks of his chosen people. The first motion is always from God, and we can never drag God under our control. And yet a kind of idolatry is always present in religious culture. We want to bring God under our control. That is the kind of attitude many believers have, and it reveals precisely that kind of idolatry.

There is another part of this idolatry, which can only be understood if we see the idol that is associated with modern science. This is the idolatry of making God "explanation". We bring God in to try to explain things that we cannot otherwise explain. "How did the universe begin?", "How did we come to be?" and all such questions. We latch onto God, especially if we do not feel that we have a good and reasonable scientific explanation. He is brought in as the Great God of the Gaps.

The Idolatry of Science

But this kind of idolatry of God as explanation will only be understood after we talk about modern science. If for purposes of illustration we speak only of the Jewish/Christian tradition, the roots of religious belief reach to some thousands of years before Christ with the prophet Abraham. However, modern science cannot properly be dated before the 16th or 17th century, roughly from the time of Galileo and then through many others to Newton. One might even wish to go back to the beginnings of the experimental method with Roger Bacon and others in the 13th century. But, at any rate, the modern science that speaks to religion today was born much later than the religion to which it speaks. It has to be recognized that the religious tradition is historically much longer and to a certain extent has that richness of the past that modern science does not.

It is a well-established historical fact that at the precise time at which modern science was being born, all of the great scientific figures were religious believers. That just happens to be the case. Newton, Leibniz, Descartes, Mersenne, Galileo, all of these and more, were religious believers. Without exception, although to some more than to others, the astounding success of the new scientific method was a great temptation to them. They were tempted to establish the foundations for religious belief with the same kind of rational certainty that they had with respect to scientific results. Many see the roots of modern atheism in precisely this kind of extreme rationalistic approach to religious belief. Trying to establish religious belief on the same sort of firm rational basis as scientific results are founded is a temptation that is always present.

The idolatry of science is, therefore, two-fold: we drag science in, certainly in Western society, to try to establish the basis of religious belief on purely rationalistic grounds. More than that, and more generally, science lures religious people of all kinds to see God first of all as "explanation" and then only secondarily to come to worship and honor him as the faithful God of Moses.

It is common to our Western mentality to make an idol of the scientific method. Some scientists and many non-scientists think that scientists know everything and that science is the only way to true and certain knowledge. That is not Science but rather it is Scientism, since it makes science a kind of god. That is idolatry. And yet any practicing scientist knows that we always struggle to come to an understanding by using the hypothetical deductive method. We collect data, we go back to our models, we revise our models, we do more computer calculations, we gather more data, and we find that they do not quite fit. In this fashion we are always struggling to come to a more complete and more certain knowledge. But we do not have it. We do not possess the truth; we hope that we are making our way towards the truth.

If we do not recognize the idolatry that is a constant factor in both science and religion, both in the past and today, then when science and religion talk to one another, all that will occur is a lot of noise, or, to recall the initial phrase with which this discussion began, the sacred cows will only be mooing.

Knowing God through Science

Several approaches can be taken to the science-religion dialogue. Here is one approach that I would like to share with you. I truly believe that God is a person and revealed himself personally to us, to his chosen people and by means of his chosen people to all of us. He did that in history, in Church traditions and in Scripture. This is not the occasion to talk about this in any detail, but there are certainly good solid foundations for believing that God is revealing himself to us in Scripture. It is certainly very firm in the Christian traditions that God also reveals himself in everything he made in Creation: in personal creation and in objective non-personal creation.

The technical approach to the attempt to probe this self-revelation of God in creation is called "analogy". It comes from the Thomistic Scholastic tradition and it refers to a relationship of similitudes, or of things that are similar. For instance, God is perfect love, and you can compare that with other kinds of love that you witness, such as the love of a mother for her child, or the faithful, long-standing love of a husband and wife for one another in a stable marriage. You see these human loves and you say: "That must be something like God's love". However, you also see imperfections in human love. There are squabbles, jealousies, infidelity, and you have to deny that those are present in God's love. That is the use of analogy. The only knowledge we can have of God, except for those who have had mystical experiences, is indirect and through analogy. If that is the case, and if God does wish to tell us about himself, then he is doing so through his creation. It follows, therefore, that I as a scientist and as a religious believer should try to use my science to see what it has to say about the God that I believe in?

Now, please notice the process that I am describing. I have never come to believe in God, nor do I think anyone has come to believe in God, by proving God's existence through anything like a scientific process. God is not found as the conclusion of a rational process. So the path that I am taking is that I believe in God because God gave himself to me. If that is the case, why should I not use my best knowledge of science to try to get an idea of what God is like? It will be only a glimmer, a shadow, but it is the one thing that I have to go on.

The Modern Science of Human Origins

It is arguably difficult to find a more divisive topic for discussion than that concerning the origins of the universe, and especially of life and of intelligence, and whether such origins can be understand without evoking a Creator God. Responses range from the extremes of a Stephen Hawking or a Pope Pius XII to almost all conceivable intermediate positions. Hawking claims that, if his quantum cosmological theory of the origins of the universe without boundary conditions is correct, then we have no need of God. To quote Hawking:

> So long as the universe had a beginning, we could suppose it had a creator. But if the universe is really completely self-contained, having no boundary or edge, it would have neither beginning nor end: it would simply be. What place, then, for a creator?[3]

Pius XII attempted to claim that with Big Bang cosmologies scientists were coming to discover what had already been known from the Book of Genesis, namely that the universe had a beginning in God's creative action. To quote Pius XII:

> ...contemporary science with one sweep back across the centuries has succeeded in bearing witness to the august instant of the primordial Fiat Lux, when along with matter there burst forth

> from nothing a sea of light and radiation... Thus, with that concreteness which is characteristic of physical proofs, modern science has confirmed the contingency of the Universe and also the well-founded deduction to the epoch when the world came forth from the hands of the Creator.[4]

Modern science, I believe, confutes both of these positions and provides for another one. Let us take one solid conclusion of modern science: we humans have come to be as products of an evolving universe and we are evolving with it. How precisely did we humans come to be in this immense evolving universe of galaxies and stars? It is quite clear that we do not know everything about this process. But it would be scientifically absurd to deny that the human brain is a result of a process of chemical complexification in an evolving universe. After the universe became rich in certain basic chemicals, those chemicals combined in successive steps to make ever more complex molecules. Finally, in some extraordinary chemical process, the human brain came to be, and it is the most complicated machine that we know.

Let us pause for a moment to review the degree of certainty that we can place in the above scenario. We certainly do not have the scientific knowledge to say how each living creature came to be in detail. We do not know precisely how each more complex chemical system came to contribute to the process of self-organization that brought about the diversity of life forms as we know them today. Most importantly, we do not know with scientific accuracy the sufficient elements in nature to have brought about the unbroken geneological continuity in evolution that we propose actually happened. There are, in brief, a number of epistemological gaps that prevent natural science from saying that a detailed theory of biotic evolution has been proven. What we have is the most adequate account conceivable at this time considering the available empirical data. That empirical data, with respect to the issue of biotic evolution, comes from various independent scientific enterprises, including molecular biology, paleontology and comparative anatomy.

Did we come about by chance or by necessity in this evolving universe? The first thing that must be said is that the problem is not formulated correctly. It is not just a question of chance or necessity because, first of all, it is both. Furthermore, there is a third element here that is very important. It is what might be called "opportunity" and it is based upon our scientific knowledge of the universe. The universe is so prolific in offering the opportunity for the success of both chance and necessary processes that such a character of the universe must be included in the discussion.

The universe is 12 to 15 billion years old, it contains about 100 billion galaxies each of which contains about 200 billion stars of an immense variety. For as many as 15 billion years the universe has been playing at the lottery. What do I mean by the lottery? When we speak about chance we mean that it is very unlikely that a certain event would happen. The "very unlikely" can be calculated in mathematical terms. Such a calculation takes into account how big the universe is, how many stars there are, how many stars would have developed planets, etc. In other words, it is not just guesswork. There is a foundation in fact for making each successive calculation. From a strictly mathematical analysis of this process, called the mathematics of nonlinear dynamics, one can say that as it goes on and more complex molecules develop, and there is more and more direction to this process.

We might further illustrate what opportunity means in the following way. Einstein once said that God does not play at dice. He was referring specifically to quantum mechanics, but his statement can be applied in general to his view of the universe. For him God made a universe to work according to established laws. This is referred to as a Newtonian Universe. It is like a clock that just keeps ticking away once you supply it energy. There are many scientists, especially evolutionary biologists, who challenge this point of view. They claim that God does play at dice because he is certain to win. That is, he made a universe prolific with opportunities for life. The point being made is that, whether or not you believe that God made it, the universe is so prolific with the possibilities for these processes to have success that we have to take the nature of the universe into consideration when we talk about how we came to be.

Considering all of the conditions necessary for life, some as yet perhaps unknown, it is difficult to determine how probable it was that it would come to be. We do know that in a very prolific universe with so many opportunities there is a narrowing down of the evolutionary process due to the nature of physics, chemistry, biology and non-linear dynamics. In other words, there has occurred a kind of intrinsic destiny towards human beings.

Implications for Religious Belief

How are we to interpret this scientific picture in terms of religious belief. Do we need God to explain this? Very succinctly my answer is no. In fact, to need God would be a very denial of God. God is not the response to a need. One gets the impression from certain religious believers that they fondly hope for the durability of certain gaps in our scientific knowledge of evolution, so that they can fill them with God. This is the exact opposite of what human intelligence is all about. We should be seeking for the fullness of God in creation. We should not need God; we should accept him when he comes to us.

The scientific picture traced above deals with the questions of origins, of how what we observe and experience today came to be. Specifically, I am asking how human beings came to be in an existing universe. The question of creation, and therefore of a God Creator, responds to the question of why there is anything in existence. Creation is not one of the ways whereby things originated, as opposed to other ways that can be thought of, including quantum cosmology and evolutionary biology. The claim that all things are created is a metaphysical and a religious claim that all that exists depends for its existence on God. It says nothing scientifically of how things came to be. On the other hand, beautiful stories are told in the Book of Genesis, that elaborate on the dependence of all things for their existence upon God.

We have referred above to the richness of religious tradition deriving from its antiquity. Here is an example of that richness. The biblical account of creation in the Book of Genesis highlights the comment made by God after each act of creation: "And he saw that it was good". The Hebrew word that is used to express "good" has, in fact, a strong indication of something aesthetically pleasing, so that, without betraying the original, one might translate the comment: "And he saw that it was beautiful". Thus, every creative act of God becomes a source of beauty and, as a sharing in God's creation, every human creation, the invention of a new scientific theory, for instance, is also a source of beauty.

A study of the Old Testament shows that the first reflection of the Jewish people was that the universe was the source of their praise of the Lord who had freed them from bondage and had chosen them as his people. The Book of Psalms, written for the most part well before the Book of Genesis, bears witness to this: "The mountains and valleys skip with joy to praise the Lord"; "The heavens reveal the glory of the Lord and the firmament proclaims his handiwork". But if these creatures of the universe were to praise the Lord, they must be good and beautiful. Upon reflecting on their goodness and beauty, God's chosen people came to realize that these creatures must come from God, and so we have the stories in Genesis in which at the end of each day God declares that what he had created is good (beautiful). The stories of Genesis are, therefore, more about God than they are about the universe and its beginning. They are not, in the first place, speaking of the origins of the created world. They are speaking of the beauty of the created world and the source of that beauty, God. The universe sings God's praises because it is beautiful; it is beautiful because God made it. In these simple affirmations we may even trace the roots of modern science in the west. The beauty of the universe invites us to know more about it and this search for knowledge leads to the discovery of a rationality innate in the universe.

But if we confront what we know of origins scientifically with religious faith in God the Creator, in the senses described above, what results? I would claim that the detailed scientific understanding of human origins in an existing universe has no bearing whatsoever on whether God exists or not.

It has a great deal to do with my knowledge of God through analogy, should I happen to believe he exists. Let me explain.

Take two rather extreme scientific views of human origins: that of Stephen Gould, who describes an episodic, totally contingent and, therefore, non-repeatable evolutionary process[5] as contrasted to a convergent evolutionary process such as that of Christian de Duve, in which the interplay of chance, necessity and opportunity leads inevitably to life and intelligence.[6] In either case, it is scientifically tenable to maintain an autonomy and self-sufficiency of the natural processes in a natural world, so that recourse to God to explain the origins of human existence is not required. It is not a question of chance in nature, which excludes God, or destiny in nature, which requires God. In neither case is God actually required.

However, if I believe in God then what nature through analogy tells me about God in one case is very different from what nature tells me about God in the other. Please note that I am not calling upon faith to adjudicate between contrasting these scientific viewpoints. I do think that convergent evolution is more consistent with God's revelation of himself in the Book of Scripture, so that, as Galileo was fond of stating, the Book of Scripture and the Book of Nature speak of the same God.

When science detects such directedness in the evolution of life in the universe, it inevitably leads us to talk about purpose in some fashion. The fear among scientists is that in talking about purpose we are inevitably going to bring God into the picture. That is not true. We do not need God to explain the universe as we see it today. But once I believe in God, the universe as I see it today says a great deal about that God.

God of the Universe and of Us

If we take the results of modern science seriously, it is difficult to believe that God is omnipotent and omniscient in the sense of the scholastic philosophers. Science tells us of a God who must be very different from God as seen by the medieval philosophers and theologians. Let us ask the hard question. Could, for instance, God after a billion years in a fifteen billion year old universe have predicted that human life would come to be? Let us suppose that God possessed the theory of everything, knew all the laws of physics, all the fundamental forces. Even then could God know with certainty that human life would come to be? If we truly accept the scientific view that, in addition to necessary processes and the immense opportunities offered by the universe, there are also chance processes, then it would appear that not even God could know the outcome with certainty. God cannot know what is not knowable. The theologian's answer would be that God is transcendent and outside of space and time. The notion of prediction clearly implies a time line. God sees everything simultaneously, if you wish. But God is also immanent and I stress the immanence in order to emphasize the uncertainties involved in human origins from within the universe itself.

This is not to place a limitation upon God. Far from it. It reveals a God who made a universe that has within it a certain dynamism and thus participates in the very creativity of God. Such a view of creation can be found in early Christian writings, especially in those of St. Augustine in his comments on Genesis. If they respect the results of modern science, religious believers must move away from the notion of a dictator God, a Newtonian God who made the universe as a watch that ticks along regularly and in which humans came to be by force of God's constant intervention.

Perhaps God should be seen more as a parent. Scripture is very rich in this thought. Scripture presents, even anthropomorphically, a God who gets angry, who disciplines, a God who nurtures the universe. Theologians already possess the concept of God's continuous creation. I think to explore modern science with this notion of continuous creation would be a very enriching experience for theologians and religious believers. God is working with the universe. The universe has a certain vitality of its own like a child does. You discipline a child but you try to preserve and enrich the individual character of the child and its own passion for life. A parent must allow the child to grow into adulthood, to come to make its own choices, to go on its own way in life. In such ways does God deal with the universe.

These are very weak images; but how else do we talk about God? We can only come to know God by analogy. The universe as we know it today through science is one way to derive analogical knowledge of God. For those who believe that modern science does indeed say something to us about God. It provides us with a challenge, an enriching challenge, to traditional beliefs about God.

Concluding Remarks: The Unification of Our Knowledge

Among the several criteria for judging the veracity of a scientific model, I would suggest including its unifying explanatory power. The inability, at times, to provide a strictly scientific explanation to what are strictly scientific problems may be an invitation to consider that the explanation lies in a teleological consideration. It is important to emphasis the word "invitation", so as to preserve the epistemological independence of the various disciplines. One is perfectly free to accept the invitation or not. One can stay firmly put within one's own discipline and continue to seek the answer there, uncontaminated by possible solutions arising elsewhere. But it seems to me that the invitation is a very real one and well founded; it, therefore, also seems to me that it requires serious reasons to reject it. Those serious reasons must confront the very long history of philosophical and theological thought that there is a person at the source of the existence of the universe and that this same person had a purpose or a design in "creating" the universe, a design which included, and is perhaps even centered upon, our existence.

The criterion for unification of our knowledge appears to extend the epistemological nature of the natural sciences towards the realm of other disciplines, such as philosophy and theology. Put in very simple terms this criterion is nothing else than a call for the unification of our knowledge. One could hardly be opposed to that. The problem arises with the application of this criterion. When is the unification not truly unifying but rather just an adulteration of knowledge obtained by one discipline with the presuppositions inherent in another discipline. History is full of examples of such adulterations. We have mentioned above the views of Stephen Hawking and of Pope Pius XII. It is for this reason that scientists have always hesitated to make use of this criterion. And yet, if applied cautiously, it appears to me to be a most creative one for the advancement of our knowledge.

The supposition is that there is a universal basis for our understanding and, since that basis cannot be self-contradictory, the understanding we have from one discipline should complement that which we have from all other disciplines. One is most faithful to one's own discipline, be it the natural sciences, the social sciences, philosophy, literature, theology, etc., if one accepts this universal basis. This means in practice that, while remaining faithful to the strict truth criteria af one's own discipline, we are open to accept the truth value of the conclusions of other disciplines. And this acceptance must not only be passive, in the sense that we do not deny those conclusions, but also active, in the sense that we integrate those conclusions into the conclusions derived from one's own proper discipline. This, of course, does not mean that there will be no conflict, even contradictions, between conclusions reached by various disciplines. But if one truly accepts the universal basis that I have spoken of above, then those conflicts and contradictions must be seen as temporary and apparent. They themselves can serve as a spur to further knowledge, since the attempt to resolve the differences will undoubtedly bring us to a richer unified understanding.

NOTES

[1] See, for example, *Wayfarers in the Cosmos: The Human Quest for Meaning,* G. V. Coyne, S. J. and A. Omizzolo (New York: The Crossroad Publishing Company, 2002).

[2] See, for example, the entry "Yahweh", by R. T. A. Murphy in the *New Catholic Encylopedia* (New York: McGraw-Hill, 1967) Vol. XIV, p. 1065.

[3] S. W. Hawking, *A Brief History of Time from the Big Bang to Black Holes* (New York: Bantam Books, 1988) p. 198.

[4] Discourses of the Popes from Pius XI to John Paul II to the Pontifical Academy of Sciences (Vatican City State: Pontificia Academia Scientiarum, 1986) Scripta Varia 66, p. 82.

[5] S. J. Gould, *Wonderful Life: The Burgess Shale and the Nature of Reality* (New York: W. W. Norton and Co., 1989).

[6] C. de Duve, *Vital Dust* (New York: Basic Books, 1995) and more recently *Life Evolving: Molecules, Mind, and Meaning* (Oxford: Oxford University Press, 2002).

International Symposium on Astrophysics Research and on the Dialogue between Science and Religion, Vatican Observatory, 2003
C. D. Impey & C. E. Petry, editors

Dare a Scientist Believe in Design?

Owen Gingerich
Harvard University

On Cone Snails and Venom

Conus cedonulli is, literally, the "I yield to none" cone. In the eighteenth century this handsomely patterned shell became the most celebrated and sought-after molluscan rarity. Two specimens were known in Europe in the early 1700s, one of which became the prize the King of Portugal's collection. In 1796 the other was auctioned for 243 guilders at a sale in which Vermeer's masterpiece, "Woman in Blue Reading a Letter", fetched a mere 43 guilders!

Cone shells, *cedonulli* among them, are considered among the most "advanced" molluscs because their anatomy includes a toxic harpoon that can spring out of the apex end of the shell. In some species the sting can be deadly even to humans.

Our first reaction upon hearing about the cone shells may well be: what wonderful design! And we may be even more impressed and probably puzzled to learn that the exquisite pattern on the shell is, during the animal's lifetime, covered by an opaque periostracum, rendering the pattern virtually invisible and therefore perplexingly useless either for survival or sexual attraction. To think in terms of deliberate design is an almost intuitive response, yet such thoughts have become strangely taboo in contemporary scientific circles. *Conus cedonulli* thus becomes a jumping off place for consideration of the question, "Are there intimations of design in the universe?" and a related and perhaps more daunting query, "Dare a scientist believe in design?"

Consider what happened when a report on studies of the mollusc toxins appeared in *Science* magazine (along with an illustration of both *Conus cedonulli* and the Vermeer painting).[1] A supplementary news article, entitled "Science Digests the Secrets of Voracious Killer Snails" remarked that "the great diversity and specificity of toxins in the venoms of the cone snails are due to the intense evolutionary pressure on the snails to stop their prey quickly, since they can't chase it down".[2]

Very promptly a letter to the editor objected that this language implied that some real pressure was driving the snails to develop the toxins:

> The reality is that those snails that produced toxins that immobilized their prey quickly tended to obtain food more often than those possessing slower-acting or no toxins, and thus over time the population of cone shells became dominated by those pos-

> sessing the fast-acting agents. There was no pressure! In the vernacular, "If it works, it works; if it don't, it don't".[3]

The response shows clearly the current philosophical orthodoxy about the non-directed nature of evolution. It also typifies the enormous change of view that has occurred over the past century with respect to the wonders of the biological world.

Design and Technology

What is now seen as the zigzag, largely accidental path to amazing organisms with astonishing adaptations was in earlier times routinely interpreted as the design of an intelligent Creator. The long neck of the giraffe, which so well adapts the creature to an environment where food is available high off the ground, would have been seen, in William Paley's words, as a "mark of contrivance, in proof of design, and of a designing Creator".[4] "Who gave white bears and white wolves to the snowy regions of the North, and as food for the bears the whale, and for the wolves, birds' eggs?" asked Johannes Kepler two centuries earlier.[5] "Great is our Lord and great his virtue and of his wisdom there is no number!" he exclaims in answer, "Use every sense for perceiving your Creator".

Even Jean Jacques Rousseau, not best known as a theist, declared, "It is impossible for me to conceive that a system of beings can be so wisely regulated without the existence of some intelligent cause which affects such regulation… I believe, therefore, that the world is governed by a wise and powerful Will".[6]

The notion of design suggests, of course, the existence of a goal-directed or end-directed process, what can aptly be termed teleology. Ernst Mayr, a leading evolutionist who has written very clearly on the modern philosophy of evolution, wisely remarks that it is futile to attempt to clarify the concept of teleology without discriminating between different types of end-directed processes. There are some kinds of inanimate natural processes that do have an end point, for example, the fall of a stone or the cooling of a heated piece of metal. There are also goal-directed processes in genetically controlled organisms:

> The third category, organic adeptness, is not directed toward an end but rather an adaptation to the environment in the widest sense of the word, acquired during evolution, largely guided by natural selection. The fourth, teleology, the cosmic one, with a purpose and predetermined goal ascribed to everything in nature, is not supported by scientific evidence.[7]

So much then, for a role for the Creator in modern biology. G. G. Simpson, writing in a more visceral fashion, declared, "Man was not the goal of evolution, which evidently had no goal. He was not planned, in an operation wholly planless".[8]

The Universe from a Dot of Energy

Yet, despite the articulate denials of cosmic teleology by the leading evolutionists of our age, there still remain enough astonishing details of the natural order to evoke a powerful feeling of awe. In our own age, science has opened up truly remarkable vistas. In the microscopic world of the atom there are marvels to stagger the imagination. At the other end of the scale, astronomers plumb the world of the large, delineating our Milky Way as a giant pinwheel galaxy containing over 200 billion stars – roughly 35 for every man, woman, and child on earth. And beyond our own stellar system, countless other galaxies are scattered out to the fringes of the universe, roughly 14 billion light-years away.

These are discoveries of the 20th century, some of them scarcely fifty years old. Yet nothing is quite as astonishing as the scientific scenario that has now been outlined for the first moments of creation. During the past few decades, knowledge of the world of the smallest possible sizes, the domain of particle physics, has been combined with astronomy to describe the universe in its opening stages. The physics ultimately fails as the nucleo-cosmologists push their calculations back to Time Zero, but they get pretty close to the beginning, to 10^{-43} second. At that point, at a second split so fine that no clock could measure it, the entire observable universe is compressed within the wavelike blur described by the uncertainty principle, so tiny and compact that it could pass through the eye of a needle. Not just this room, or the earth, or the solar system, but the entire universe squeezed into a dense dot of pure energy. And then comes the explosion. "There is no way to express that explosion" writes the poet Robinson Jeffers:

> ...All that exists
> Roars into flame, the tortured fragments rush away from
> each other into all the sky, new universes
> Jewel the black breast of night; and far off the outer nebulae
> like charging spearmen again
> Invade emptiness.[9]

It's an amazing picture, of pure and incredibly energetic light being transformed into matter, and leaving its vestiges behind. It's even more astonishing when we realize that the final fate of the universe, whether it will expand forever or fall back on itself to a future Big Crunch, was determined in that opening moment.

Now the paragraphs you've just read were written just over a decade ago, and today we know there are a couple of things wrong with them. First, Robinson Jeffers' poem says that the nebulae like charging spearmen *again* invade emptiness. Jeffers' younger brother, Hamilton, was an astronomer at Lick Observatory, and from his brother he may have learned about the oncepopular cosmological theory of an oscillating universe, in which a universe could have formed over and over again. Today the observational evidence favors a universe that will expand forever, so the oscillating model has fallen into disfavor.

Second, that dense dot of pure energy that could pass through the eye of a needle brings up a very interesting problem. How does the dot know when to explode? In 10^{-40} second light cannot travel very far, not even across an atom, much less across the eye of a needle. There is no way for one side of the dot to know when the other side started to expand! An intriguing solution to this conundrum was suggested two decades ago, named by Alan Guth, one of its inventors, the "inflation" scenario. In this sub-theme to the Big Bang, at a very early instant, just as the gravitational force separated out from the other basic forces of the universe, for a moment gravitation acted repulsively and the space expanded by billions of billions of billions of times – actually even more than this – with the result that the entire part of the universe now visible to us was once exceedingly tiny and entirely connected at the speed of light.

The Universe on a Knife Edge

Yet, despite the articulate denials of cosmic teleology by the leading evolutionists of our age, there still remain enough astonishing details of the natural order to evoke a feeling of awe – so much so that cosmologists have even given it a name: the anthropic principle. The discussion arose originally when some physicists noticed that even small variations in some of the constants of nature would have led to a universe in which life could not exist. For example, had the original energy of the Big Bang explosion been less, the universe would have fallen back onto itself long before there had been time to build the elements required for life and to produce from them intelligent, sentient beings. Had the energy been more, it is quite possible that the density would have dropped too swiftly for stars and galaxies to form. The balance between the energy of expansion and the gravitational braking had to be right to 1 part in 10^{59} – an incredible ratio. These and many other details were so extraordinarily right that it seemed the universe had been expressly designed for humankind.

Today we would say that this incredible balance was not an accident, or, one might argue, direct evidence of God's designing hand in the opening moment of the universe, not what Aristotle might have called a "fact-initself", but now a "reasoned fact", a phenomenon with an explanation, for this balance must inevitably follow as a consequence of that split second of enormous inflation. Does that mean that this evidence for God's being and presence in our universe has gone away? A "fact-in-itself" might call for God's fine-tuning hand in the history of the universe, while the march of science has negated the necessity of such a role. But a "reasoned fact" still resonates with the magnificence of God's original designs that have made such a universe possible. It is the difference between contingent events and fundamental planning for the nature of the universe itself. Let me return presently to this idea, and to the anthropic principle, the principle that the universe is somehow extraordinarily, and in fact necessarily, congenial for the origin of intelligent life.

On Carbon and Resonance

One of the first scientists to consider how the environment itself made life possible was the Harvard chemist L. J. Henderson. In 1913, after Darwin's emphasis on the fitness of organisms for their various environments, Henderson wrote a fascinating book entitled *The Fitness of the Environment*, which pointed out that the organisms themselves would not exist except for certain properties of matter. He argued for the uniqueness of carbon as the chemical basis of life, and everything we have learned since then, from the nature of the hydrogen bond to the structure of DNA, reinforces his argument. But today it is possible to go still further and to probe the origin of carbon itself, through its synthesis deep inside evolving stars.

Carbon is the fourth most common atom in our galaxy, after hydrogen, helium, and oxygen. A carbon nucleus can be made by merging three helium nuclei, but a triple collision is tolerably rare. It would be easier if two helium nuclei would stick together to form beryllium, but beryllium is not very stable. Nevertheless, sometimes before the two helium nuclei can come unstuck, a third helium nucleus strikes home, and a carbon nucleus results. And here the details of the internal energy levels of the carbon nucleus become interesting: it turns out that there is precisely the right resonance within the carbon that helps this process along.

Let me digress a bit to remind you about resonance. You've no doubt heard that opera singers such as Enrico Caruso could shatter a wine glass by singing just the right note with enough volume. I don't doubt the story, because in the lectures at our Science Center at Harvard, about half a dozen wine glasses are shattered each year using sound waves. It's necessary to tune the audio generator through the frequency spectrum to just the right note where the glass begins to vibrate – the specific resonance for that particular goblet – and then to turn up the volume so that the glass vibrates more and more violently until it flies apart.

The specific resonances within atomic nuclei are something like that, except in this case the particular energy enables the parts to stick together rather than to fly apart. In the carbon atom, the resonance just happens to match the combined energy of the beryllium atom and a colliding helium nucleus. Without it, there would be relatively few carbon atoms. Similarly, the internal details of the oxygen nucleus play a critical role. Oxygen can be formed by combining helium and carbon nuclei, but the corresponding resonance level in the oxygen nucleus is half a percent too low for the combination to stay together easily. Had the resonance level in the carbon been 4% lower, there would be essentially no carbon. Had that level in the oxygen been only half a percent higher, virtually all of the carbon would have been converted to oxygen. Without that carbon abundance, neither you nor I would be here now.

I am told that the late Fred Hoyle, who together with Willy Fowler found this remarkable nuclear arrangement, said that nothing shook his atheism as much as this discovery. Occasionally Fred Hoyle and I sat down to discuss one or another astronomical or historical point, but I never had enough nerve to ask him if his atheism had really been shaken by finding the nuclear resonance structure of carbon and oxygen. However, the answer came rather clearly in the November 1981 issue of the Cal Tech alumni magazine, where he wrote:

> Would you not say to yourself, "Some supercalculating intellect must have designed the properties of the carbon atom, otherwise the chance of my finding such an atom through the blind forces of nature would be utterly minuscule". Of course you would... A common sense interpretation of the facts suggests that a superintellect has monkeyed with physics, as well as with chemistry and biology, and that there are no blind forces worth speaking about in nature. The numbers one calculates from the facts seem to me so overwhelming as to put this conclusion almost beyond question.[10]

For me, it is not a matter of proofs and demonstrations, but of making sense of the astonishing cosmic order that the sciences repeatedly reveal. Fred Hoyle and I differed on lots of questions, but on this we agreed: a common sense and satisfying interpretation of our world suggests the designing hand of a super-intelligence.

Design and Biology

Impressive as the evidences of design in the astrophysical world may be, however, I personally find even more remarkable those from the biological realm. As Walt Whitman proclaimed, "A leaf of grass is no less than the journey work of the stars".[11] I would go still farther and assert that stellar evolution is child's play compared to the complexity of DNA in grass or mice. Whitman goes on, musing that:

> ...the tree toad is a chef-d'oeuvre for the highest,
> And the running blackberry would adorn the parlors of heaven,
> And the narrowest hinge in my hand puts to scorn all machinery,
> And the cow crunching with depress'd head surpasses any statue,
> And a mouse is miracle enough to stagger sextillions of infidels.

Even Hoyle, by his allusion to the biology, seems to agree that the formation of, say, DNA, is so improbable as to require a superintelligence. Such biochemical arguments were popularized about forty years ago by Lecomte du Nouy in his book *Human Destiny.* Du Noüy estimated the probability of forming a 2000-atom protein as something like one part in 10.[321] He wrote, "Events which, even when we admit very numerous experiments, reactions, or shakings per second, need an infinitely longer time than the estimated duration of the earth in order to have one chance, on the average, to manifest themselves can, it would seem, be considered as impossible in the human sense".[12]

Du Noüy went on to say, "To study the most interesting phenomena, namely Life and eventually Man, we are, therefore, forced to call on antichance, as Eddington called it; a "cheater" who systematically violates the laws of large numbers, the statistical laws which deny any individuality to the particles considered".[13]

The game plan for evolutionary theory, however, is to find the accidental, contingent ways in which these unlikely and seemingly impossible events could have taken place. The evolutionists do not seek an automatic scheme – mechanistic in the sense that Newtonian mechanics is determined – but some random pathways whose existence could be at least partially retraced by induction from the fragmentary historical record. But when the working procedure becomes raised to a philosophy of nature, the practitioners begin to place their faith in the roulette of chance and they find that du Noüy and Hoyle are an aggravation to their assumptions about the meaninglessness of the universe.

Scientists and Their Beliefs

Despite the reluctance of many evolutionary theorists, there does seem to be enough evidence of design in the universe to give some pause. In fact, scientists who wish to deny the role of design seem to have taken over the anthropic principle. Briefly stated, they have turned the original argument on its head. Rather than accepting that we are here because of a deliberate supernatural design, they claim that the universe simply must be this way *because* we are here; had the universe been otherwise, we would not be here to observe ourselves, and that is that. As I said, I am doubtful that you can convert a skeptic by the argument of design, and the discussions of the anthropic principle seem to prove the point.

But once again I return to my central question, "Dare a scientist believe in design?" and I pause to remark on the somewhat curious status of "belief" within science. Some years ago I conducted a workshop for a rather diverse group of Christians, and I asked, "Can a theist believe in evolution?" I got a variety of responses, but it didn't occur to any of them to challenge what it might mean to *believe* in evolution. Does that mean to have faith in evolution in a religious sense? I have heard one leading paleontologist announce himself as a "devout evolutionist" when asked his faith, and I guess that is a possibility. But when pressed, most scientists would, I think, claim only that they accept evolution as a working hypothesis.

In everyday, non-philosophical usage, most people, scientists included, would say they believe in the results of science and that they believe the results of science to be true. Yet, and this is the anomalous part, most scientists would be mildly offended at the thought that their beliefs constituted an act of faith in a largely unproved but intricate system of coherencies. Actually, surprisingly little in science itself is accepted by "proof". Let's take Newtonian mechanics as an example. Newton had no proof that the Earth moved, or that the Sun was the center of the planetary system. Yet, without that assumption, his system didn't make much sense. What he had was an elaborate and highly successful scheme of both explanation and prediction, and most people had no trouble believing it, but what they were accepting as truth was a grand scheme whose validity rested on its coherency, not on any proof. Thus, when a convincing stellar parallax was measured in 1838, or when Foucault swung his famous pendulum at 2 a.m. on Wednesday morning, January 8, 1851, these supposed proofs of the revolution and of the rotation of the earth did not produce a sudden, newfound acceptance of the heliocentric cosmology. The battle had long before been won by a persuasiveness that rested not on proof but on coherency.

Now if we understand that science's great success has been in the production of a remarkably coherent view of nature rather than in an intricately dovetailed set of proofs, then I would argue that a belief in design can also have a legitimate place in human understanding even if it falls short of proof. What is needed is a consistent and coherent world view, and at least for some of us, the universe is easier to comprehend if we assume that it has both purpose and design, even if this cannot be proven with a tight logical deduction.

Nevertheless, there has been a persistent criticism that arguments from design will cause scientific investigators to give up too easily. If the resonance levels of carbon and oxygen are seen as a miracle of creation, would a Christian physicist try to understand more deeply why, from the mechanistic view of physics, the levels are that particular way and not in some other configuration? Might it not be potentially detrimental to the faith to explain a miracle? And so we come face to face with our original query: "Dare a scientist believe in design?"

Kepler

There is, I shall argue, no contradiction between holding a staunch belief in supernatural design and being a creative scientist, and perhaps no one illustrates this point better than the seventeenth-century astronomer Johannes Kepler. He was one of the most creative astronomers of all time, a man who played a major role in bringing about the acceptance of the Copernican system through the efficacy of his tables of planetary motion. Now one of the principal reasons Kepler was a Copernican arose from his deeply held belief that the sun-cantered arrangement reflected the divine design of the cosmos: The Sun at the center was the image of God, the outer surface of the star-studded heavenly sphere was the image of Christ, and the intermediate planetary space represented the Holy Spirit. These were not ephemeral notions of his student years, but a constant obsession that inspired and drove him through his entire life.

Writing to a favorite correspondent, Herwart von Hohenburg, he said, "Copernicus piously exclaimed, 'So vast, without any question, is the Divine handiwork of the Almighty Creator'. ...Yet we must not infer that bigness is of special importance; otherwise the crocodile or elephant would be closer to God's heart than man".[14] To his teacher Michael Maestlin back in Tubingen he wrote, "For a long time I wanted to be a theologian; for a long time I was restless. Now, however, behold how through my effort God is being celebrated in astronomy!"[15]

Today Kepler is best remembered for his discovery of the elliptical form of the planets' orbits. This discovery and another, the so-called law of areas, are chronicled in his *Astronomia Nova,* truly the New Astronomy. In its introduction he defended his Copernicanism from the point of view of that the heavens declare the glory of God:

> If someone is so dumb that he cannot grasp the science of astronomy, or so weak that he cannot believe Copernicus without offending his piety, I advise him to mind his own business, to quit this worldly pursuit, to stay at home and cultivate his own garden, and when he turns his eyes toward the visible heavens (the only way he sees them), let him with his whole heart pour forth praise and gratitude to God the Creator. Let him assure himself that he is serving God no less than the astronomer to whom God has granted the privilege of seeing more clearly with the eyes of the mind.[16]

Kepler's life and works provide central evidence that an individual can be both a creative scientist and a believer in divine design in the universe, and that indeed the very motivation for the scientific research can stem from a desire to trace God's handiwork.

Darwin

In the centuries that followed, many scientists took inspiration from the idea that the heavens declared the glory of God, but God's hand appeared less and less in their physical explanations. In a sense, one of the fundamental consequences of the scientific revolution, in which the ancient geocentric universe gave way to a vast heliocentric plan governed by gravitation, was the secularization of the natural world.

Darwin's theory was of a quite different sort from Newton's. He sought some fundamental explanation for patterns of similarities as well as differences within the biological kingdoms, and a way to understand the remarkable adaptation of the organisms beyond a simple attribution to God's designing hand. Darwin's explanation eventually relied on historical contingency rather than mechanical necessity. As such, his theory lacked the compelling predictive power that arises from the necessity of gravitation, or of conservation of angular momentum, or of any number of other physical laws. What it lacked in sheer predictive power it achieved in its immense explanatory power, a spectacular new coherency of understanding.

With the secularization of the physical world that followed in the wake of the scientific revolution of the sixteenth and seventeenth centuries, the community became divided between the deists, who put God outside the universe as the Spirit who set it all into motion according to physical laws, and the theists, who still maintained an active role for God within the world. This was, of course, a theological or philosophical option, not a decision required by any scientific observations of the world itself. After Darwin's evolutionary theory was raised to a philosophy, with its inherent denial of design, the apparent choice swung more sharply from deism/theism to atheism/theism. Science remained a neutral way of explaining things, neither anti-God nor atheistic. Many people were (and are) extremely uncomfortable with a way of looking at the universe that did not explicitly require the hand of God. But it did not mean the universe was actually like that, just that science generally has no other way of working.

Nevertheless, high random opportunism (as opposed to design) has been raised to such a level of scientific orthodoxy that some of our contemporaries forget that this is just a tactic of science, an assumption, and not a guaranteed principle of reality. Few, however, have enunciated the mechanistic credo so stridently as the evolutionary biologist and historian of science William B. Provine, who has recently written:

> When Darwin deduced the theory of natural selection to explain the adaptations in which he had previously seen the handiwork of God, he knew that he was committing cultural murder. He understood immediately that if natural selection explained adaptations, and evolution by descent were true, then the argument from design was dead and all that went with it, namely the existence of a personal god, free will, life after death, immutable moral laws, and ultimate meaning in life. The immediate reactions to Darwin's *On the Origin of Species* exhibit, in addition to favorable and admiring responses from a relatively few scientists, an understandable fear and disgust that has never disappeared from Western culture.[17]

Provine, in defending the gospel of meaninglessness, goes on to say that if modern evolutionary biology is true, then lofty desires such as divinely inspired moral laws and some kind of ultimate meaning in life are hopeless.

Christian Biochemistry

I'm not sure why Professor Provine has such fear and loathing of design, but apparently, despite the example of Kepler (and of Newton and many others), he is still afraid that the arguments from design may block the march of science. Such a view is perhaps not totally unfounded. Let me explain.

Several years ago I participated in a remarkable conference of theists and atheists in Dallas. One session considered the origin of life, and a group of Christian biochemists argued that the historical record was non-scientific since it was impossible to perform scientific experiments on history. Furthermore, they amassed considerable evidence that the current scenarios of the chemical evolution of life were untenable. One of the atheists aligned against them, Professor Clifford Matthews from the University of Illinois at Chicago, conceded that their criticisms had considerable validity. Calling their book on *The Mystery of Life's*

Origins[18] brilliant, he summarized their arguments with respect to the standard picture of chemical evolution as saying, "(1) the evidence is weak, (2) the premises are wrong, and (3) the whole thing is impossible". Of course, he did not accept their final conclusion that a new kind of science was required.

I soon found myself in the somewhat anomalous position that to me, the atheists' approach was much more interesting than the theists'. That particular group of Christian biochemists had concluded that ordinary science didn't work in such a historical situation, that is, with respect to the origin of life, and they attempted to delineate an alternative "origin science" in which the explicit guiding hand of God could make possible what was otherwise beyond any probability. The real reason I admired the atheist biochemists so much was that they hadn't given up. They were still proposing ingenious avenues whereby catalytic effects in the chemistry made the events far more likely. "Let us not flee to a supernaturalistic explanation", they said, "let us not retreat from the laboratory".

Now it might be that the chemistry of life's origins *are* forever beyond human comprehension, but I see no way to establish that scientifically. Therefore it seems to me to be part of science to keep trying, even if ultimately there is no accessible answer. Apparently this reasoning has some cogency, because the ringleader of the group, Charles Thaxton, at least partly backed off of this position, and today we don't hear much about origin science. But meanwhile, a new generation has reclothed some of these same ideas under the name "intelligent design". Using some of the same evidences that impress me, from both the physical and biological realms, they press the case still further and argue that some of the evolutionary steps make sense only when taken in a large bundle, a form of macroevolution that demands the explicit involvement of both a designing mind and a designer's hand. My theological presuppositions do incline me to be sympathetic to this point of view, but as a scientist I accept methodological naturalism as a research strategy.

On Amish and Six-Fingered Dwarfs

Let me digress for a moment to describe a particular case that has been intriguing me recently. As an introductory footnote, I should say that all four of my paternal great great grandfathers were Amish ministers, which enhances my fascination with this situation among the highly inbred population of Amish in Lancaster County, Pennsylvania. Among this group there is the occasional appearance of a rare pathology known as six-finger dwarfism. There are approximately 75 known cases, about half of which were still-born, and the remarkable circumstance is that in every instance both parents could trace their ancestry to a single Amish couple who immigrated to Pennsylvania around 1750. In other words, each parent carried the mutant recessive gene inherited from either the mother or the father in that original couple – there is no way, of course, to know which one carried it to America. Two years ago the single altered nucleic acid in the DNA was discovered. Just one substitution is enough to cause a host of changes including the production of a sixth finger.

I cite this case for several reasons. First, it shows how intricate and extraordinary the control mechanisms are in the DNA, and how cautious we have to be in making claims about how macroevolutionary changes can or cannot occur. Secondly, it shows just how ambiguous our interpretation of physical events must necessarily be. Was that mutation an accident? Was God surprised? It's hard to believe that it was somehow foreordained in the Big Bang itself. The fundamental integrity of atoms could, to put it in human, anthropomorphic language, have been planned and designed in the beginning, but with regard to the mutation itself, the two obvious theological choices are either that God's hand is continually at work disguised in the ambiguity of the uncertainty principle, or that the purposes of creation are general and not specific, so that God is learning as the process goes on.

Transcendence

Am I contradicting myself to say, on the one hand, that the resonance levels in carbon and oxygen point to a superintelligent design, and on the other hand, that science must continue to search for underlying reasons why the resonance levels are that way and not some other way? I think not, for even if it is shown that those levels had to be the way they are because of some fundamental, invariable reason, there is still the miracle of design that led it so, choice or not. Even if it would no longer be a "fact-in-itself", but a reasoned fact, the design would still be there. Thus, I see no reason that an appreciation of the astonishing details of design should prevent us from trying to search further into their underlying causes. Hence I'm not prepared to concede that arguments from design are necessarily contra- scientific in their nature.

Perhaps part of Provine's outrage came because he was responding to Phillip Johnson, Professor of Law at Berkeley, who is an articulate legal champion of the right to believe in God as Creator and Designer, and a critic of an evolutionary process running entirely by chance.[19] Earlier I mentioned the incredible odds calculated by Lecomte du Noüy against the chance formation of a protein molecule. Since we do have proteins, and since a mechanistic science has been highly successful, the overwhelming reaction has just been to ignore du Noüy, since he is so obviously wrong. But is he? For science to overcome the odds, it is necessary for us to postulate catalysts and unknown pathways to make the formation of life from inert matter enormously easier, and it is of course precisely such pathways that are the challenge of science to find. But is not the existence of such pathways also evidence of design? And are they not inevitable? That is what materialists such as Provine do not want to hear, but as Hoyle says, the numbers one calculates puts the matter beyond question.

So, while I worry about from those Christian biochemists who postulate some new kind of "origin science", or their successors who argue for "intelligent design", I do think a science totally devoid of the idea of design may be in danger of running into a blank wall. And this brings me to ask again, is the idea of design a threat to science? and I answer no, perhaps design might even be a necessary ingredient in science.

In reflecting on these questions I have attempted, in a somewhat guarded way, to delineate a place for design both in the world of science and in the world of theology. As Kepler once said of astrology, the stars impel, but they do not compel.[20] There is persuasion here, but no proof. However, even in the hands of secular philosophers the modern mythologies of the heavens, the beginnings and endings implied in the Big Bang, give hints of ultimate realities beyond the universe itself. Milton Munitz, in his closely argued book, *Cosmic Understanding,*[21] declares that our cosmology leads logically to the idea of a transcendence beyond time and space, giving lie to the notion that the cosmos is all there is, or was, or ever will be.

Munitz, in coming to the concept of transcendence, describes it as unknowable, which is somewhat paradoxical, since if the transcendence is unknowable then we cannot know that it is unknowable. Could the unknowable have revealed itself? Logic is defied by the idea that the unknowable might have communicated to us, but coherence is not. For me, it makes sense to suppose that the superintelligence, the transcendence, the ground of being in Paul Tillich's formulation, has revealed itself through prophets in all ages, and supremely in the life of Jesus Christ.

To believe this requires accepting teleology and purpose. And if that purpose includes contemplative intelligent life that can admire the universe and can search out its secrets, then the cosmos must have those properties congenial to life. For me, part of the coherency of the universe is that it is purposeful – though probably it takes the eyes of faith to accept that. But given that understanding, then the anthropic principle that states that our universe must be congenial to life also becomes the evidence of design. This brings to mind a few lines in Whitman's *Leaves of Grass:*

> A child said *What is the grass?* fetching it to me with full hands;
> How could I answer the child? I do not know what it is any more than he.
>
> I guess it is the handkerchief of the Lord,
> A scented gift and remembrancer designedly drops,
> Bearing the owner's name someway in the corners,
> that we may see and remark, and say *Whose?*[22]

So, just as I believe that the Book of Scripture illumines the pathway to God, I also believe that the Book of Nature, with its astonishing details – the blade of grass, the *Conus cedonulli,* or the resonance levels of the carbon atom – also suggests a God of purpose and a God of design. And I think my belief makes me no less a scientist.

To conclude, I turn once again to Kepler, who wrote:

> If I have been allured into brashness by the wonderful beauty of thy works, or if I have loved my own glory among men, while advancing in work destined for thy glory, gently and mercifully pardon me: and finally, deign graciously to cause that these demonstrations may lead to thy glory and to the salvation of souls, and nowhere be an obstacle to that. Amen.[23]

NOTES

[1] B. M. Olivera, *et al.*, "Diversity of *Conus* Neuropeptides", *Science*, 249 (20 July 1990), pp. 257-63.
[2] Marcia Barinaga, "Science Digests the Secrets of Voracious Killer Snails", *Science*, 249 (20 July 1990), pp. 250-51.
[3] James L. Carew, "'Purposeful' Evolution" (letter), *Science*, 249 (24 August 1990), p. 843.
[4] William Paley, *Natural Theology; or, Evidences of the Existence and Attributes of the Deity Collected from the Appearances of Nature* (Edinburgh, 1816), Chapter 5, section 5, p. 61.
[5] Johannes Kepler, *Harmonies of the World*, in *Great Books of the Western World*, Vol. 16 (Chicago, 1952), p. 1085.
[6] J. J. Rousseau, *Profession of Faith of a Savoyard Vicar* (1765), quoted in Alan Lightman and Owen Gingerich, "When Do Anomalies Begin?" *Science*, 225 (7 February 1992), pp. 690-95.
[7] Ernst Mayr, "The Ideological Resistance to Darwin's Theory of Natural Selection", *Proceedings of the American Philosophical Society*, 135 (1991), pp. 123-39, on. p. 131. One clause has been inserted from Mayr's Craaford Prize Lecture as edited in *Scientific American*, 283 No. 1 (July, 2000), p. 82.
[8] George Gaylord Simpson, *The Meaning of Evolution* (Mentor Edition, New York, 1951), p. 143.
[9] Robinson Jeffers, "The Great Explosion", in *The Beginning and the End and Other Poems* (New York, 1963).
[10] Fred Hoyle, "The Universe: Past and Present Reflections", in *Engineering and Science*, November, 1981, pp. 8-12, especially p. 12.
[11] Whitman, "Song of Myself", stanza 31, in *Leaves of Grass* (Boston, "1891-2 edition").
[12] Du Noüy, *Human Destiny*, (New York, 1947), p. 35.
[13] *Human Destiny*, p. 38.
[14] *Johannes Kepler Gesammelte Werke*, 13, nr. 107 (16 December 1598), pp. 144-58; my translation based on the one by Carola Baumgardt, *Johannes Kepler Life and Letters* (New York, 1951) pp. 48-49.
[15] *Johannes Kepler Gesammelte Werke*, 13, nr. 23 (3 October 1595), pp. 256-57; from Gerald Holton, "Johannes Kepler's Universe: Its Physics and Metaphysics", *American Journal of Physics*, 24 (1956), pp. 340-51, especially p. 351.
[16] Slightly abridged and modified from my translation in *Great Ideas Today 1983* (Chicago, 1983), pp. 321-22.
[17] *First Things*, No. 6, October 1990, p. 23.
[18] Charles B. Thaxton, Walter L. Bradley, and Roger L. Olsen, *The Mystery of Life's Origin: Reassessing Current Theories* (New York, 1984).
[19] See also Phillip E. Johnson, *Darwin on Trial* (Washington, D.C., 1991).
[20] Kepler, *Harmonice mundi* (Frankfurt, 1619), IV, Ch. 7.
[21] Milton K. Munitz, *Cosmic Understanding: Philosophy and Science of the Universe* (Princeton, 1986).
[22] Whitman, "Song of Myself", stanza 6, in *Leaves of Grass* (Boston, "1891-2 edition").
[23] End of Book V, Chapter 9 of *Harmonice mundi*, Johannes Kepler Gesammelte Werke, 6, p. 362; my translation is based on the ones by Charles Glenn Wallis in *Great Books of the Western World*, Vol. 16, and by Eric J. Aiton, A. M. Duncan and J. V. Field, *Memoirs of the American Philosophical Society*, 209 (Philadelphia, 1997).

International Symposium on Astrophysics Research and on the Dialogue between Science and Religion, Vatican Observatory, 2003
C. D. Impey & C. E. Petry, editors

Truth and Beauty in Cosmology

Chris Impey
University of Arizona

Prelude

Cosmology is not always beautiful to the practitioner. Consider the petty professional rivalries, the computers that can't calculate the quantity you want, the clever ideas discarded like a child's broken toys, the photons that travel billions of light years only to hit a single cloud or fall uselessly to the ground beside your telescope. On the other hand, there is an undeniable grandeur in the quest to understand the universe. For a species only a thousand generations removed from savagery, many of whose members currently live in poverty and physical misery, the goal of understanding the origin of all time, space, and matter is audacious indeed.

Successive stages of the Copernican Revolution have taught us our place in the universe. The Sun is just one of billions of stars in the Milky Way, many of which will prove to be orbited by rocks similar to the one on which we find ourselves. The Milky Way is just one of billions of galaxies flung through the vastness of expanding space. The stuff of which we are made is not even typical – our atoms are greatly outnumbered by feeble photons and by a sea of mysterious dark matter particles. Now modern cosmology teaches us that the universe itself may not be unique. Our realm of time and space may be one of many spawned by the quantum chaos of the Big Bang. Timothy Ferris has written an excellent history and overview.[1]

Is the universe beautiful? A working cosmologist might blink uncomprehendingly at the question. As George Santayana has said "Art critics talk about theories of art; artists talk about where to buy good turpentine". Yet the question is well motivated, because a similar aesthetic sense drives both cosmologists and creative thinkers in other disciplines. We will start with some quotes to frame the discussion:

> Definition: "Beauty is the proper conformity of the parts to one another and to the whole". W. *Heisenberg*
>
> Definition: "There is no excellent beauty that does not have some strangeness in the proportion". *F. Bacon*
>
> Proposition: "Mathematics is the archetype of the beautiful". *J. Kepler*
>
> Proposition: "It is vain to do with more what can be done with fewer". *W. Ockham*

Provocation: "All metaphors are imperfect, and that is the beauty of them". *R. Frost*

Provocation: "When I have to choose between the true and the beautiful, I choose the beautiful". *H. Weyl*

This article addresses truth and beauty in cosmology, with minor excursions to consider the nature of beauty, the role of mathematics, and the nature of physical reality. Since modern theories of nature unite the micro-world and the macro-world, the subject matter ranges from subatomic particles to the whole universe, with we humans caught in between as bewildered participants and observers. The term"God" is used informally, to represent organizing principles, and to acknowledge the fact that scientists cannot ascribe a cause to the universe. Perhaps the entire enterprise of cosmology represents unalloyed hubris – that judgment is left to the reader.

God Loves Structure

"To see a World in a grain of sand,
And Heaven in a wild flower;
Hold infinity in the palm of your hand,
And eternity in an hour". *W. Blake*

No law of physics would be violated if the universe were filled with an undifferentiated sea of particles and photons. Yet it is not (and of course we would not be here to ponder the situation if it were). Instead, the universe contains a riot of structure on every scale. Short-range nuclear forces act as the glue that facilitate the creation of elements in the periodic table. Longer-range electrical forces permit these atoms to be combined into molecules and compounds in myriad ways. Finally, the long arm of gravity inexorably sculpts the universe, forming objects that range from asteroids, the size of a small town, to superclusters of galaxies, which light would take 300 million light years to traverse.

The laws of physics are simple, yet the world is complex. Physical scientists tend to be reductionists, but that urge should be resisted, because reductionism is doomed to failure. It is just not feasible to compute the quantum mechanical state of a complex molecule, or a network diagram for the electrical activity in the brain, or the gravitational interactions of all the stars in the universe. Knowledge of the microscopic components does not imply complete understanding of the behavior of the macroscopic whole. New insights are required.

One of the most important insights was due to the mathematician Benoit Mandlebrot who demonstrated the role of fractals in nature.[2] His classic example was the coastline of Norway. If you take a map of Norway and examine smaller and smaller sections of coastline, the total length continues to increase because there is successively more structure on smaller scales. The long arc of the coastline is serrated with fiords, and each fiord contains many inlets, and so on. This is also true of vertical geographic features. In any mountain range, there will be a few towering summits, a larger number of moderate peaks, and many foothills or smaller features. Any geometric shape with this character is a fractal.

The mathematical description of fractal behavior is a power law. If the logarithm of the number of features on a particular scale is plotted against the logarithm of the size scale, the result is a straight line that slopes downward – indicating more features, or "power", on smaller and smaller scales. If power laws were confined to coastlines and mountain ranges, they would be nothing more than a mathematical curiosity. But over the last 40 years, power law behavior has been observed in an amazing array of physical situations.

On the Earth, in addition to the examples just given, the distribution of the thickness of sedimentary layers is a power law, as is the distribution of earthquake strengths, and the number of extinctions throughout the history of life. Beyond the Earth, the strength of quasar light variations is a power law, as is the distribution of X-rays from solar flares, and the distribution of pulsar glitches caused by quakes in the crust of a neutron star.[3] The distribution of large-scale structure in the universe is also a power law, which provides a simple description of the filaments, walls and voids in the galaxy distribution.[4] Matter in the universe has a fractal dimension of about 1.7, partway between stringy structures (1D) and sheet-like structures (2D). On the very largest scales, the universe is smooth, as exemplified by the microwave background radiation.

More mundane terrestrial phenomena also display power law behavior. The distribution of word frequency in the English language is a power law, as is the distribution of city sizes, and amount of time different cars will wait in a traffic delay. The physicist Per Bak has presented these intriguing cases, and he has used the simple example of a sand pile to gain another insight into the nature of complexity. As grains of sand are added to a conical pile of sand, the overall shape is maintained through a series of miniature avalanches and slippages. The total system maintains a regular simplicity even though the triggering events are random and chaotic. The statistical behavior of individual random events leads to structure and coherence.

A simple mathematical form captures many of the properties of cooperative phenomena. The recursion relation, $\chi_{n+1} = \lambda\chi_n\,(1-\chi_n)$ couples events that are adjacent in time or in space. When the parameter lambda is small, the system is static. For slightly larger values, cyclic or periodic behavior is seen. When lambda is very large, the behavior of the system is chaotic and unpredictable. There is usually a high degree of structure and complexity at the transition between cyclic and chaotic regimes.[5] This description has proved useful in understanding situations as diverse as a set of linked oscillators, or the growth of a crystal, or the electrical network in the brain.

The ubiquity of power laws in nature is an indication of cooperative phenomena at work. For example, power law size distributions result from the complementary processes of attrition (sand at a beach, the asteroid belt) and accretion (formation of a raindrop, or a planet). In earthquakes and sand piles, sudden change occurs when small scales couple to large scales. The technical term for this situation self-organized criticality, but Blake's poetic allusion turns out to be highly apt.

God Loves Disorder

"All change distributes energy,
spills what cannot be gathered again.
Each meal, each smile – scatters treasure, lets fall
gold straws woven from the resurgent dust". *J. Updike*

The structure in the universe occurs within a larger framework of disorder. One measure of this is the enormous number of photons in the universe. There are nearly a hundred million photons for every particle, and the radiation field is completely unstructured. These photons are currently imperceptible to us, because they have been redshifted by the expansion of the universe to feeble microwaves. However, project the expansion back towards the Big Bang and these photons were much more energetic. About 300,000 years after the Big Bang, the photons were visible and the entire universe had the temperature of the surface of a star like the Sun. About 10,000 years after the Big Bang, the photons had X-ray energies and their violent collisions prevented any gravitational structure from forming.

Another manifestation of disorder in the universe is entropy. Technically, entropy is a measure of the number of possible states of a system, best illustrated with the analogy of a deck of cards. Imagine a deck with the cards all ordered in rank and suit. There is only one way for the deck to be sequenced this way. Now imagine shuffling the deck. The color and rank sequences will rapidly be broken up. There are many ways a deck can contain 8 or 9 cards in sequence, and vastly many more where it can contain two or three in sequence. This increase in disorder with time is a characteristic of all physical systems. Equivalently, a system will always tend to arrange itself in the lowest energy configuration.

Think of mismatched socks in your drawer, or the many pieces of a broken wine glass, or the complete mixing that takes place when you stir cream into your coffee. In fact, the "arrow of time" is a reflection of the possible number of microscopic states of a system. The laws of microscopic physics are invariant to time, but in a statistical sense, the ensemble shows a unidirectional behavior. It is relatively easy to match up all the loose socks in your drawer, and somewhat more work to reassemble the broken wine glass. But no amount of shuffling will produce a completely ordered deck of cards, and no amount of stirring will separate the cream from the coffee.

The universe itself also follows the trend of increasing disorder. Regardless of the outcome of the cosmic expansion, overall entropy will increase. If the average density of matter exceeds a critical value, then the cosmic expansion will eventually be overcome by gravity and the universe will collapse. This collapse will proceed like a mirror image of the original expansion. All the structure that has been painstakingly created by gravity will be obliterated in the heat death of the Big Crunch.[6]

However, it appears almost certain that the universe will expand forever. Even when the ubiquitous dark matter is added in, there is insufficient matter for gravity to overcome the expansion. As a result, the galaxies will continue to sail apart. Within them, the cycle of star death will eventually be broken as more mass is locked up in white dwarfs, neutron stars and black holes, and less gas is available to form a new generation of stars. Over many billions of years, each galaxy will slowly fade from view. Finally, if current theories of particle physics are correct, protons will decay and matter will "dissolve". The ultimate state of the universe will then be a cold and featureless sea of electrons, positrons and low energy photons.

Entropy is thus a powerful physical principle that applies on microscopic and cosmic scales. The great physicist Ludwig Boltzmann was sufficiently impressed that he had the mathematical definition of entropy inscribed on his tombstone. Sir Arthur Eddington was also unequivocal as to the centrality of the concept:

> The law that entropy increases holds, I think, the supreme position among the laws of Nature. If someone points out that your pet theory of the Universe disagrees with Maxwell's equations – then so much the worse for Maxwell's equations. If it is contradicted by observation – well, these experimentalists do bungle sometimes. But if your theory is found to be against the second Law of Thermodynamics – I can give you no hope, there is nothing for it but to collapse in deepest humiliation.[7]

The Nature of Beauty

So far, we have avoided the issue of how beauty is defined in science. We might start by noting that some aspects of beauty are hardwired, like the attraction felt by a newborn baby for their mother's face. The fact that the universe contains structure also elicits aesthetic appreciation. Physical laws have imprinted patterns in nature that are pleasing to humans – from the reflective planes of a tiny crystal, to the soaring peaks of a mountain range, to the majesty of a pinwheel galaxy set against the dark velvet of night. It is less clear that beauty can be found from the inchoate disorder of entropy. However, one of the surprises of modern science is the fact that behavior can arise in a complex system that is not entirely deducible from the initial conditions. Patterns can emerge from chaos.

It is important not to be too parochial in defining beauty. The quote by Heisenberg at the beginning of the article applies equally to Science and to Arts. Proportion and harmony are features of creative works as diverse as Shakespeare's *King Lear,* Beethoven's *Missa Solemnis,* and Joyce's *Ulysses.* The aesthetic does not have to be classical – the same is true for a play by Tom Stoppard, or a Beatles ballad, or a novel by Martin Amis. Beauty in proportion and form is scale-free; it can be found equally well in a haiku or an epic poem, a short story or a novel, or in a writer's life work.[8] With modern creative arts, perhaps the key aesthetic element is "truthtelling". Regardless of how abstract or personalized the work is, it succeeds by creating a resonance with the viewer and often by making a larger connection to the human condition.

While artists do not all agree on the definition of beauty, they might agree on the recognition of beauty based on the emotional response that it elicits. Scientists have an equally well-formed sense of beauty, but its basis might not be recognizable to those outside the field. The author J. W. N. Sullivan, who wrote influential biographies of Newton and Beethoven, has written:

> Since the primary object of the scientific theory is to express the harmonies which are found to exist in nature, we see at once that these theories must have an aesthetic value. The measure of the success of a scientific theory is, in fact, a measure of its aesthetic value, since it is a measure of the extent to which it has introduced harmony in what was before chaos.[9]

The astrophysicist Subramanyan Chandrasekhar believed that Einstein's general theory of relativity is the most beautiful creation of human thought.[10] Einstein himself wrote at the end of his first paper on the subject: "Scarcely anyone who fully comprehends the theory can escape from its magic". The insight of the great scientist carries with it the same flavor of revelation as the creation of an artist. Listen to Heisenberg describing his discovery of the rules of quantum mechanics:

> It was almost three o'clock in the morning before the final results of my computation lay before me. The energy principle had held for all terms, and I could no longer doubt the mathematical coherence and consistency of the kind of quantum mechanics to which my calculations points. At first, I was deeply alarmed. I had the feeling that, through the surface of atomic phenomena, I was looking at a strangely beautiful interior, and felt almost giddy at the thought that I now had to probe this wealth of mathematical structure nature had so generously spread out before me. I was too excited to sleep.[11]

By now it should be clear that mathematics lies at the heart of the scientific aesthetic. This can be true quite literally, as noted by Boltzmann "Even as a musician can recognize his Mozart, Beethoven, or Schubert after hearing a first few bars, so can a mathematician recognize his Cauchy, Gauss, Jacobi, Helmholtz, or Kirchhoff after the first few pages". At least the issue of truth becomes moot – mathematics is not useful unless it is also true. However, the central role of mathematics in scientific knowledge deserves further exploration.

The Nature of Mathematics

From Pythagoras and Plato, we have inherited the idea of the universe as a mathematical entity. The essence of science is the discovery of patterns in nature, and Plato was well aware that these patterns are only imperfectly manifested in the everyday world. Plato knew that the material world could not be understood applying mathematics, hence the inscription carved above the door of his Academy: “Let None But Geometers Enter Here”. With Pythagoras, mathematics attained cult status. Spurred on by the discovery of the rules of musical harmony, Pythagoras elevated harmony to an organizing principle of the cosmos. Two thousand years later, Kepler seized on the “harmony of the spheres” with true mystical fervour as he applied it to the orbits in the solar system.

Without any doubt, mathematics is one of the highest human achievements. Until a baby is about three months old, an object removed from view is essentially removed from the infant universe. Then the brain evolves to the point where an object can be retained in memory after it is removed from view. This event is the birth of abstraction, the basis of mathematics. Modern mathematics is not confined to intellectual esoterica; it is a wide-ranging activity with a vital role in the modern economy. For the most part, scientists have devised new mathematical techniques to solve practical problems. However, a venerable question still presents itself: Is mathematics invented or discovered?

Some insights are gained by examining the worlds of inner and outer space. In the late 1970s, Mandlebrot explored the properties of a simple recursive set based on complex numbers, numbers that contain a real and imaginary part. The rule he used to form a set of numbers is amazingly simple: z ’ $z^2 + c$, where c is a constant, and z is a complex number of the form $a + ib$. The real and imaginary parts of the number combine to make a map. When Mandlebrot first explored his set, he thought his computer was malfunctioning, because he saw a rich pattern of loops and filaments, protuberances and whorls. Amazingly, when he zoomed in on a small region, an equally rich web of structure was revealed. On and on it went, apparently without limit. As Roger Penrose pointed out: “The Mandlebrot set is not an invention of the human mind: it was a discovery. Like Mount Everest, the Mandlebrot set is just there!”[12]

The Pythagorean theorem, $a^2 + b^2 = c^2$, is an algebraic statement of Euclidean geometry, it applies only to spaces that are linear and congruent. In the 19th century, mathematicians developed the formalism of curved space and space with an arbitrary number of dimensions. It was pure, abstract mathematics – nobody had any idea that it might apply to the real world. Early in the 20th century, Einstein developed his new theory of gravity. General relativity makes a profound connection between the material contents of the universe and the curvature of space. As the physicist John Wheeler has put it: “Matter tells space how to curve, and space tells matter how to move”. The global curvature of space is extremely subtle, but the distortion of space by matter has been observed near black holes and in the phenomenon of gravitational lensing.

Most theorists in physics and cosmology subscribe to a Platonic view of the universe. They make a distinction between mathematics that is derived to solve particular problems, and mathematics that encapsulates, and expands our understanding of, a basic feature of the physical universe. Cosmology has experienced a century of rapid progress using a gravity theory based on geometry and tensor mathematics. However, our understanding of the universe is still far from complete, and we do not know how far mathematics will take us.

God Loves Chance

"An optimist proclaims
that we live in the best of all possible worlds;
and the pessimist fears
this is true". *J. Cabell*

Is the universe beautiful when it is governed primarily by chance? Human experience is strongly rooted in ideas of determinacy and causality. We move through the world as if events have identifiable causes, even if we may not always be clever enough to identify them. Although we accept that measurements are always imprecise, we are confident that we can drive down the error to an arbitrarily small level with sufficient data, or with better apparatus. This certainty melts away when we have to consider the interior world of the atom.

The quantum revolution early in the 20th century led not only to a new way of doing physics, but also to a new way of looking at the world. Imagine you are in a darkened room trying to locate an object. You might shine a beam of light at it, and the reflected light would reveal its position. But what if the object is very small, perhaps not much larger than an atom? Since light carries energy, as it interacts with the object and returns to you, it will transfer energy to the object and give it motion. So you have located the object but at the expense of unknown motion. At the limit of a single atom, even if your light beam shrinks to a single photon, it will influence the state of the atom. No amount of cunning will help you – lowering the energy of the photon will reduce the energy it gives to the particle, but the wavelength becomes larger so the position sensing is more imprecise.

Werner Heisenberg enshrined this concept as his uncertainty principle, and the subatomic world makes no sense without it. It is better called an indeterminacy principle, since it does not reflect any limitation in the measuring equipment. Rather, it is a fundamental veil that is drawn over our knowledge of the physical world. Imagine you measure the position of a star in the sky. Multiple measurements give a scatter of positions, and scientists normally adopt the mean as the best estimate of the true position and the scatter in the data as a measure of observational uncertainty. By the central limit theorem, more measurements give you a better estimate of the true position (or you could refine your technique to give a smaller observation error). The situation with a subatomic particle could not be more different. There is no "true" position, and the location of the particle can only be described by a probability distribution.

Thus, there are no absolutes in the quantum world. There is only a continual trade-off between knowledge of position and motion, with determinacy limited by Planck's constant, a very tiny number. Heisenberg presented a second form of his uncertainty principle that expresses a tradeoff between knowledge of energy and time. So it turns out that the conservation of energy, an iron clad principle of the macro-world, can be violated in the micro-world. Energy can be created from nothing, if only for a brief instant. Manifestations of both versions of the uncertainty principle are observed routinely in physics labs.

With discomfort, we face the role that metaphor plays in science. Waves and particles are macroscopic phenomena, and those words conjure up concrete imagery that does not map perfectly down to the quantum world. Physicists learn to trust the mathematical formalism of wave functions and probabilities, because it works (most electronics and much of our hi-tech world depends on it). We could not understand how the Sun shines without quantum physics. Protons carry a positive charge that should act as an impenetrable wall of electric force to prevent them coming together. Yet there is a small but finite probability that the proton will breach the wall, and this quantum tunnelling must be included to explain the rate of proton fusion in the Sun and other stars.

It gets worse. Our idea of causality is rooted in the idea that light is an absolute speed limit and that nothing can be in two places at one time. Yet a series of experiments over the past twenty years have shown that when two photons travel through an apparatus, the state of one of the photons can instantaneously affect the state of the other. This example of "non-local" physics appears to imply communications travelling faster than the speed of light. The formalism of quantum mechanics adequately describes these experiments, but only by introducing negative probabilities into the calculation of the wave function.[13]

Philosophers, and many physicists, have reacted with dismay to this situation. In addition to Einstein's famous admonition about God not playing dice with the universe, there have been hopes of a deeper set of "hidden variables" to make quantum effects comprehensible. Another ruse has been to postulate that each quantum state spawns a new universe with that particular outcome – the "many worlds" interpretation. This extravagant idea was explored by Olaf Stapledon in the influential science fiction work *Star Maker*:

> Whenever a creature was faced with several possible courses of action, it took them all, thereby creating many distinct histories of the cosmos. Since in every evolutionary sequence of the cosmos there were many creatures and each was constantly faced with many possible courses, and the combinations of their courses were innumerable, and infinity of distinct universes exfoliated from every moment of every temporal sequence in this cosmos.[14]

These alternatives seem worse than the problem they were designed to address. Why invent a new and untested underlying theory when quantum mechanics works so well? Perhaps we should accept that our intuition breaks down at the level of a single photon or a single particle, and that we cannot apply classical expectations to the words wave, particle, field, and coherence. Why conjure up an infinite array of parallel universes? It does nothing to explain the mysteries of the quantum world and it violates Ockham's razor – the dictum of simplicity – egregiously. We might take solace in the fact that Planck's constant is so small that quantum graininess is not apparent in the everyday world. Macroscopic objects are real and we know where they are, and events have causes that precede them, thanks to thermodynamics and the ensemble behavior of particles. However, quantum effects play out in two extraordinary ways in the cosmos as a whole.

Space is observed to be smooth and geometrically flat. This is an unlikely outcome in the standard Big Bang model, because the early universe was warped and filled with topological defects, or rifts in space-time. Cosmologists believe that the universe went through a rapid inflation, from the size of a proton to the size of a grapefruit, a merest 10^{-35} seconds after the Big.[15] Inflation ensures that space now will be smooth and flat – imagine a small and curved balloon that is inflated so much that any section of the material appears to be flat. Inflation also implies that the region of space that is within the light grasp of our telescopes is only a tiny fraction of the whole amount, because space was expanding much faster than the speed of light in this first instant.[16] The physical universe must be much larger than the observable universe. Lastly, tiny quantum fluctuations were amplified by inflation to a size large enough to become the seeds for structure formation. The Big Bang means that galaxies are our historical windows onto the quantum world.

The second quantum effect is equally bizarre. Astronomers talk about the vacuum of space, but to a physicist there is no true vacuum. The lowest energy state of space is not zero but a quantum of energy above zero. This energy, with its attendant ability to create particle and anti-particle pairs, creates pressure that counters the universe attraction of gravity. Einstein coopted this pressure force to balance gravity and match his cosmological model to what he had been told was a static universe – his self-avowed "great blunder". Standard Big Bang models always decelerate, because gravity acts as a foot on the brake. However, recent observations of distant supernovae indicate that the universe is accelerating, which can only be explained by the action of vacuum energy. (There is the additional minor embarrassment that physics predicts a vacuum energy 1048 times larger than is observed in the universe.) So quantum effects seem to be driving the expansion of the universe.

God Loves Symmetry

"You boil it in sawdust, you salt it in glue;
You condense it with locusts in tape;
Still keeping one principle objective in view –
To preserve its symmetrical shape". *L. Carroll*

Science often consists of looking for patterns in nature. Ancient astronomers looked for predictable motions in the heavens. Kepler turned these into laws of planetary motion, which are explained by the single gravity law of Newton. Ancient alchemists noticed patterns in the behavior of the elements. Mendeleev arranged the pattern into a primitive version of the periodic table, which is explained by way electrons change their state within the atom.

The largest insights occur when apparently disparate phenomena are united through a new theory. Faraday suspected a linkage between the current flowing through a wire and the spinning of a compass needle; then Maxwell revealed the symmetry between electricity and magnetism with an elegant set of equations. Fifty years ago, Hermann Weyl demonstrated that each conservation law in physics is based on an underlying symmetry in nature. His quote at the beginning of the article indicates that he used an aesthetic sense of symmetry as his guide to the truth.[17]

Symmetry is esteemed in both art and music. Whether it is the design and proportion of the Great Masters, or the contrapuntal forms of Bach, or the layout of an Elizabethan sonnet, or the Palladian style of architecture, balance and symmetry are firmly established as classical virtues. Modern art forms have loosened these strictures, but they have not entirely discarded them. Symmetry is desirable but not if it is absolute; perfect symmetry is usually sterile in the arts. Thus we might be most attracted by the presence of geometric forms as long as they are modulated by the imperfections of the real world. Think of the slight flaw that renders an otherwise beautiful face irresistible.

We live in a universe of badly broken symmetry. There are four fundamental forces – gravity, electromagetism, the weak force that is responsible for radioactive decay, and the strong force that holds the atomic nucleus together. In the everyday world, these forces are all distinct, with very different strengths and ranges. Thirty years ago, theorists united the electromagnetic and weak nuclear forces, and accelerators have probed the energies at which these two forces are indeed observed to melt together. The next hurdle in the quest for unification is truly formidable. The strong nuclear force only joins the fold at energies far beyond current or projected technology.

If accelerators fail us, the universe can be our laboratory. In the earliest instants after the Big Bang, the temperature was so high that particles and antiparticles were liberally created in equal numbers from pure radiant energy. There was no structure or form, just a chaotic ocean of violent interactions. All the forces except gravity had the same strength. Then as the universe expanded and cooled, the symmetry was slightly broken. The result was a tiny excess of particles over antiparticles. When the particles and antiparticles annihilated to form photons, one in a billion particles was left over, like jilted lovers. The photons, dimmed and cooled by billions of years of cosmic expansion, are seen as the microwave background radiation. The particles have been moulded by gravity into 50 billion galaxies containing 10^{20} stars. If the symmetry had been perfect, radiation would rule totally and we would not be here.

Gravity is like a chimera just beyond our reach. In the ultimate expression of unification, gravity joins the other three forces in a superforce that explains both the micro-and the macro-world. Nobody is sure of the final form of such a theory, but it is likely to be based on the geometry of ten-dimensional space-time.[18] String theory replaces families of particles with the oscillations and undulations of e- dimensional "branes". As physicist Michiko Kaku has noted: "Of all physical theories, string theory unites by far the largest number of branches of mathematics into a single coherent picture". The austere beauty of the mathematics persuades many scientists that string theory must be right.

The Nature of Reality

What is real? Physicist John Wheeler has said that "What we call 'reality' is in large measure a papier mache construction, an immense labour of imagination and theory filled in between the iron posts of observation". Our knowledge is not based on shifting sands. We have learned to predict and control the natural world, as evidenced by the pervasive power of technology in our lives. Yet, on the smallest and largest scales, metaphors and analogies fail and we take refuge in the mathematical form of the theories. Like the party game where you trust a stranger as you fall back into their arms, the mathematics has caught us every time.

The use of imaginary numbers is a good example. The ancient Greeks had already ventured far from the safe terrain of natural numbers with their discovery of irrational and transcendental numbers; a student of Pythagoras was put to death for experimenting with irrationals. But by accepting the use of the square root of a negative number as a tool, we can attack algebraic problems that cannot be solved in any other way. Thus emboldened, mathematicians added imaginary numbers to their toolkit. We have seen how they opened the door to the baroque world of the Mandlebrot set. Early last century, Paul Dirac found symmetric solutions to the equations of the quantum theory. Rather than discard the solution that involved the imaginary i, he embraced it as a mirror world called antimatter. Twentynine years later, antiparticles were created in the laboratory for the first time.

This faith in mathematics has been rewarded by theories of successively greater scope and power. High-energy physicists are closing in the Higgs field, a mathematical entity that is presumed to account for the fact that particles have mass. String theorists are conjuring up the ultimate symmetry, where familiar forms of matter and energy are shadowy reflections of an abstract world of strings and membranes.

It takes a visionary and provocateur like John Wheeler to remind us that we are a part of this process. He has described a game of twenty questions with a trick played on the questioner: no answer has been selected when the questioner returns to the room. Rather, the players answer each question to be consistent with the answers to previous questions. The symbolism of the story:

> The world, we once believed, exists "out there" independent of any act of observation. I, entering the room, thought the room contained a definite word. In actuality the word was developed step by step through the questions I raised, as information about the electron is brought into being by experiments that the observer chooses to make. Had I asked different questions or the same questions in a different order I would have ended up with a different word as the experimenter would have ended up with a

> different story for the doings of the electron... In the game no word is a word until that word is promoted to reality by the choice of questions asked and answers given.[19]

The universe itself may be subject to whim. The limit of the Big Bang theory is the Planck era, a mere 10^{-43} seconds after creation, when the entirety of space was curved and knotted and tiny as a subatomic particle. One version of the theory called chaotic eternal inflation postulates that universes are continually spawned from this quantum foam. In each universe the physical laws are different, subject to the random nature of quantum fluctuations. Some take flight, like ours, and become large and old and filled with structures carved by gravity. Others might exist only for a moment before being crushed out of existence. Yet others may have different physical properties – universes with equal amounts of matter and antimatter, universes where gravity rivals the force that binds the atom, high entropy universes where time runs backwards. Welcome to the multiverse.

Ockham's razor has been discarded once again. Cosmologists are motivated to speculate about this proliferation of universes by the peculiar properties of this one. If the strength of any force or constant of nature were different, no law of physics would be broken, yet the universe would have properties that would render our form of life impossible. Carbon chemistry needs time, space, stable atoms and stable stars. The universe seems to have properties that are "just so" for our existence, an idea called the cosmological anthropic principle.[20] In its weak form, the anthropic principle is no more than a tautology – we can only observe a universe that has properties that would allow us to exist. However, the strong form raises the existence of life to a guiding principle.

Is this good science or sophistry? The anthropic principle does not obey the precepts of science by making clear and testable predictions, and it may use an unduly restrictive definition of life. Carbon chemistry may not be the only way to code information and transmit complexity. If selforganization can emerge from other forms of matter and energy, then the range of potential universes that could accommodate life might be much larger than we imagine. With speculation rampant, some commentators have accused physicists and cosmologists of "ironic" science, accounting for the inexplicable with the unobservable.[21] Martin Rees admits as much:

> "The multiverse is an unobservable theoretical construct whose manifestation helps us account for the way the world is". However, the modern history of science has shown us that audacity can often lead to insight.

Coda

Like Robinson Crusoe, we find ourselves on an island in uncharted seas. We do not know how many times life has started on planets around the 10^{20} stars in the universe, nor do we know how many times creatures have emerged to ponder their role in the universe. In the gulfs of time and space that we have explored with telescopes, humans are insignificant and science seems frail. Perhaps we are psychologically wired to seek order and beauty. In *The Waves,* Virginia Woolf wrote:

> There is a square; there is an oblong. The players take the square and place it on the oblong. They place it very accurately; they make a perfect dwelling place. Very little is left outside. The structure is now visible; what was inchoate is here stated; we are not so various or mean; we have made oblongs and placed them upon squares. This is our triumph; this is our consolation.[22]

Three hundred years ago, Leibnitz wrote "The first question that should be asked, would be why there is something rather than nothing? For nothing is simpler than something". Cosmology has started to address the deepest questions of our existence. At this point in the journey, we have glimpsed order that emerges from chaos, organizing principles that are mathematical and abstract, a universe whose origin may have been a chance quantum event. The universe is a gift; how could it not be beautiful?

NOTES

[1] T. Ferris, *Coming of Age in the Milky Way* (New York: Morrow and Company, 1988).
[2] B. Mandlebrot, *The Fractal Geometry of Nature* (New York: Freeman, 1983).
[3] P. Bak, *How Nature Works* (New York: Springer-Verlag, 1996).
[4] P. J. E. Peebles, *Physical Cosmology* (Princeton: Princeton University Press, 1993).
[5] S. A. Kauffmann, *The Origins of Order* (Oxford: Oxford University Press, 1993).
[6] M. J. Rees, *Before the Beginning* (Reading, Mass.: Perseus, 1997).
[7] A. Eddington, *The Nature of the Physical World* (New York: Macmillan, 1948).
[8] P. Medawar and J. Shelley, *Structure in Science and Art* (Amsterdam: Excerpta Medica, 1980).
[9] J. W. N. Sullivan, quoted by S. Chandrasekhar in *Truth and Beauty: Aesthetics and Motivations in Science* (Chicago: University of Chicago Press, 1987).
[10] S. Chandrasekhar, *Truth and Beauty: Aesthetics and Motivations in Science* (Chicago: University of Chicago Press, 1987).
[11] W. Heisenberg, *Physics and Beyond: Encounters and Conversations* (New York: Harper and Row, 1958).
[12] R. Penrose, *The Emperor's New Mind* (Oxford: Oxford University Press, 1989).
[13] T. Rothman and G. Sudarshan, *Doubt and Certainty* (Reading, Mass.: Perseus, 1998).
[14] O. Stapledon, *Star Maker* (London: Penguin, 1972).
[15] A. H. Guth and A. P. Lightman, *The Inflationary Universe* (New York: Perseus, 1998).
[16] T. Ferris, *The Whole Shebang* (New York: Simon and Schuster, 1997).
[17] F. Wilczek and B. Devine, *Longing for the Harmonies* (New York: Norton, 1988).
[18] B. Greene, *The Elegant Universe* (New York: Vintage, 2000).
[19] J. A. Wheeler, *Geons, Black Holes, and Quantum Foam* (New York: Norton, 1998).
[20] J. D. Barrow and F. Tipler, *The Cosmological Anthropic Principle* (Oxford: Oxford University Press, 1986).
[21] J. Horgan, *End of Science* (New York: Little, Brown, 1997).
[22] V. Woolf, *The Waves* (New York: Penguin, 1931).

International Symposium on Astrophysics Research and on the Dialogue between Science and Religion, Vatican Observatory, 2003
C. D. Impey & C. E. Petry, editors

Anthropic Reasoning in Cosmology

Ernan McMullin

University of Notre Dame

Introduction

Ever since the mid-1970s there has been much talk both of "anthropic principles" and "anthropic explanations" in discussions of early universe cosmology. Before reviewing the present state of a rather convoluted debate, it may be worthwhile to point out, first, that "anthropic" forms of explanation are routine in a number of the sciences already, those sciences where human agency is at issue, and second, that debate about the property of "anthropic" reference in cosmology had already begun before the 1970s.

Anthropic Explanation in the Sciences

Deep in an excavation in, say, northern Siberia, an archaeologist finds an animal bone bearing odd markings. Can these be the product of human agency? Or were they brought about by some process to which the bone was subject over the years, a process not involving human intervention? Questions of this general sort are commonplace in archaeology, physical anthropology and paleontology.

Sometimes a question like this can be easily settled: such markings can be recognized as the effect of a combination of weathering processes or of the work of animal predators. Or the markings may carry the unmistakable stamp of human intelligence: either direct, a clear symbolic intent, or indirect, the product of human-made tools employed in butchering. Often, the issue may be difficult to resolve and controversy swirls around it. The imprint of intelligence is not always decisive: was this stone deliberately shaped as a hand-axe, or might it have taken on this shape through the action of natural forces of abrasion? On a larger scale, are the apparently anomalous climatic phenomena that threaten human well-being in many parts of the world today the effect, in part at least, of climate warming – the cumulative effect of human agency? Anthropic explanation of global change becomes ever more evident and ever more disturbing.

In discussions of this sort, rival scenarios have to be considered and their likelihoods estimated. Context will be important. How long has the bone been in the ground in this particular location? To what sorts of process would it have been subject? What is the likelihood of the presence of human agents at the time the bone was deposited? Suppose that no evidence of humans in that area at that time has so far been unearthed. Suppose, further, that there is strong reason to question whether humans made their way to that area until centuries later. The more unlikely it is on contextual grounds that the markings on the bone *could* testify to human presence there at that time, the stronger the direct case would have to be that these markings *do* show definite signs of human agency if that case is to be taken seriously.

The point here is a simple one whose relevance to our topic will be apparent later on. To establish an anthropic explanation in cases like this, one has to scrutinize not only the likelihood of non-anthropic alternatives but also all relevant aspects of the anthropic postulate itself: how likely in terms of antecedent would it be to find an anthropic cause in this context? One does not proceed through a probability "filter", first definitively excluding the non-anthropic alternatives and concluding, in consequence, to the anthropic one. The intrinsic likelihood of anthropic agency of this sort in this context must also be taken into account. If a bone is found in an undisturbed location that dates back several million years, one has to look kindly at the non-anthropic possibilities in explaining the pattern of markings on it no matter how improbable they seem, as well as, of course, looking critically at the steps involved in making the age estimate for the bone.

Anthropic Intimations in Cosmology: Early

Cosmology in something of the modern manner might be said to have begun, with Descartes' *Le Monde* or its influential earlier summary in his *Discourse on the Method.* There was nothing anthropic about the "chaos" of particles in motion from which the universe began, according to his proposal. Yet from that chaos, according to him, all the physical complexity of our universe would later develop merely by the operation of the laws of mechanics over an unspecified (but obviously very great) period of time. To bring about that complexity of a star, a planet, animal life, even a human body, no constraint would need to be set on the initial cosmic conditions according to him. A "chaos" would do: his cosmology thus implicitly embodied what I have elsewhere called the "principle of indifference".[1]

Reaction to Descartes' cosmology tended to be negative, particularly in England where Boyle, Ray, and others pointed to the implausibility of accounting in this manner for the clear evidences of means-end adaptation in the living world. But it was Newton who most strongly urged the inadequacy of the Cartesian proposal in the domain of cosmology. Not only was the Cartesian mechanism of vortices "pressed with many difficulties" as he argued in the General Scholium added to the second edition of the *Principia,* it simply could not account for the known complexity of the planetary system. But, more important, a "chaos" left to itself never could give rise to the order and stability of the system we know: "This most beautiful system of the sun, planets, and comets, could only proceed from the counsel and dominion of an intelligent and powerful Being".[2]

In his letters to his disciple, Richard Bentley, Newton had already spelled out the kind of shaping that would be needed. How to make the Sun and planets form in an infinite space merely by the operation of gravity already posed insoluble problems, problems that could not be solved by "mere natural causes".[3] Placing a sun "to warm and enlighten all the rest" at the planetary center must have been due to the "Author of the System". The planets' motions and placement likewise could only have come about through the operation of "a Cause which understood and compared together" the precise quantities of matter, the distances of each planet from the sun, and the velocities each should have. The initial parameters of the system would thus have to be very severely constrained in order that a stable planetary complex of the kind we know could originate. And only an allpowerful and above all *intelligent* cause (very well skilled in mechanics and geometry!) could have brought this about.

This was a direct challenge to the Cartesian principle of indifference. Instead, Newton proposed a teleological alternative where the Creator's intelligence grasps what constraints have to be placed on the initial cosmic conditions to enable the sort of cosmos to develop that the Creator would have been likely to desire. Given the explicit reference to the Creator's purposes, the choice and imposition of these constraints could well be called "fine-tuning", a metaphor of which more will be said later. His motivation was clearly to construct a persuasive natural theology, in response to critics of his new mechanics who held that the self-aufficient science of the *Principia* had, effectively, banished God from His universe.

Was this an anthropic explanation? Not primarily – Newton's emphasis in the General Scholium is on God's power and goodness and desire for order. Yet there is a hint that this order would also serve human purposes, that without the order initially imposed on the planetary cosmos, human life could not have flourished. It was Richard Bentley, however, who developed the anthropic dimension of the initial ordering imposed on the planetary system – without such ordering, the world would, he argues, have been "unfit for the Divine purposes in creating vegetable and sensitive and rational creatures".[4] But even for Bentley, it was the "Confutation of Atheism" that the title of his Boyle Lectures promised. The emphasis on the fine-tuning needed in the initial setting of the universe was mainly to bring out that in the Newtonian, by contrast with the Cartesian, cosmology there was direct testimony to the need for the Creator's purposive shaping of the initial cosmic conditions. Still, one can say, that in Newton's own construal of his cosmology, there is at least a hint of an anthropic explanation: an anthropic motive on the Creator's part in the fine-tuning to which Newton believed his science could testify.

The problem with fine-tuning is that the next generation of scientists takes it as a challenge and tries to show it was not really needed; that with some imaginative reworking, the resources of an unconstrained system as cosmic origin turn out to be aufficient to yield in time the desired complexities. Newton urgently *wanted* his system to be insufficient without the interposition of a Creator's intelligence. But his successors did not share his urgency in that regard. In 1755, a young admirer of Newton, Immanuel Kant, showed how an original nebulous mass could contract into a rotating disc which under further gravitational contraction could yield not only our planetary system but explain why all the planets revolve in the same direction and roughly in the same plane as well as giving the first inkling of what the Milky Way might really be. Newton's fine-tuning claims and the natural theology built on them gradually collapsed, with a further push from Laplace along the way.

Although Kant had refused to follow his idol, Newton, in one anthropic direction, he more than compensated for this by propounding perhaps the boldest "anthropic" transformation of natural science ever conceived, and it was in large part prompted by the exalted epistemic status that Newton's mechanics had already achieved and the need to find an adequate foundation for that status. The story is a familiar one: Kant made space and time transcendentally ideal, the product of human ways of knowing, not the basic features of an independent real world his predecessors had unanimously supposed.[5] Were human subjects to be removed, he claims, "all the relations of objects in space and time, indeed space and time themselves, would vanish; as appearances, they cannot exist in themselves, but only in us".[6] The orderliness of Newtonian science was to be rooted in the cognitive abilities of human knowers.

This was an anthropic principle of the most sweeping kind imaginable. From it Kant went on to construct an immensely complex philosophical system, one that purported to provide the requisite metaphysical foundations for the natural sciences generally, including cosmology. I do not intend to follow this thread here; my reason for recalling it is only to underline an easily overlooked fact: the phrase "anthropic principle" may seem to be of very recent vintage, but the notion itself, in a far more radical form than anything so far dreamt of recent cosmology, is long familiar in philosophy of science.

Anthropic Intimations in Cosmology: Later

The unexpected turn taken by cosmology in the twentieth century was mainly due to the combined force of three advances. First was Einstein's General Theory of Relativity that offered alternatives to the infinities of Newtonian space and time and showed that space, time, and matter are themselves intimately interlinked. There could now be an equation, startlingly, for the universe as a whole. Even more startling, de Sitter and Friedman showed that the equations of General Relativity could have solutions requiring the universe itself to expand or contract. The second advance was Hubble's discovery of the redshift of the galaxies, most easily interpreted as indicating that all were moving away from one another. From that it was only a short step for Georges Lemaître to bring these two radical notions together to imagine an expanding universe satisfying Einstein's equations, with its constituent galactic elements moving away from each other due to the expansion of space itself.

One of those who did most to popularize this new cosmological model was Arthur Eddington, who had done distinguished work on relativity theory and on stellar evolution. But his neo-Kantian sympathies led him in directions that few would later follow. All we know of the world, he argued, is the structure we infer from pointer-reading coincidences. The most general features of that structure, notably the pure numbers expressing the basic physical constants as ratios, are determined by the nature of observation, measurement, and ultimately by the nature of the human mind itself. A purely epistemological analysis of these features, aided by ingenious and complicated deployment of group theory, enabled him (he claimed) to calculate the numerical values (necessarily integers) of the four primary physical constants, expressed as pure numbers: the ratio of the proton/electron masses, the fine-structure constant, the ratio of the electrical to the gravitational force, and a number that could be identified, he says, as (among other things) the number of particles in the universe.[7] This last number, the "cosmical number" as he called it, is not for him contingent, as one might have expected; it is decided by "the influence of the sensory equipment with which we observe and the intellectual equipment with which we formulate the results of observation as knowledge".[8] Or again, once we decide "that the right way to find out about the universe is to measure things, we are committed to an analytical conception which implicitly divided the universe into 3/2 x 136 x 2^{256} particles".[9] More generally, "All the fundamental laws and constants of physics can be deduced unambiguously from *a priori* considerations and are therefore wholly subjective".[10]

In Eddington's philosophy of science, man really does become the measure of all things, or at least all things measurable. The idea that measures like the fine-structure constant could have been predicted in advance did not sit well with Eddington's fellow physicists, needless to say, and when that constant turned out to have a fractional value and not the exact integral 137 that Eddington had extracted from his group-theoretic construction, his colleagues were not slow to make the discrepancy known. The discovery of particles other than protons and electrons made the notion of the "number of particles in the universe" even more problematic than Eddington himself had to some extent recognized. But it was the criticism from philosophers that was sharpest.[11] The idea that the values of the basic physical constants are necessarily what they are in the light of some more encompassing theory was at least conceivable. But that these values could be derived *a priori* from so exiguous a starting point as our human ways of knowing the physical world was so vulnerable to challenge that the debate about his work more or less died with Eddington himself in 1944. His was, undoubtedly, the most ambitious version of an "anthropic principle" in modern cosmology.

Interest in the supposed large-number coincidences to which Eddington had drawn attention continued, however.[12] One contribution, in particular, excited a good deal of attention. Impressed by Eddington's emphasis on the significance of the dimensionless numbers but unsympathetic with his derivation of these from a purely internal analysis of epistemology, Paul Dirac focused the discussion in a different direction. What struck him most was the presence among the puzzling numbers of some very large ones, and the further fact that these took the form $(10^{40})^n$, with n = 1 or 2. Numbers of the order of unity like the fine-structure ratio, 137, and the ratio of the proton to the electron masses, 1840, somehow seemed to be more or less at the expected scale. But why these very *large* numbers? (The numbers were all ratios in order to express them in dimensionless form). The three he pointed to were the third and fourth of Eddington's numbers: the ratio of the electric and gravitational forces between, say, proton and electron (~10^{39}), the mass of the universe (expressed as a ratio between the universe's mass and that of a proton, hence equivalently, the number of particles in the universe, ~10^{80}), and a third that was Dirac's own special concern, the "Hubble age" of the universe (the ratio of its age to the time taken by light to cross an atom, ~10^{39}). Believing that these "large number coincidences" required explanation, his hypothesis was that there was an invariant relationship between these numbers because of some as yet unknown law of nature.

Though this obviously did not *explain* the relationship, it postulated that it was not merely coincidental, that there was an underlying reason of some sort for it. And this *did* allow for an explanation of the other puzzling feature of the three numbers, their large size. One of the numbers, the Hubble age, changes over time. Since the universe is now very old, the number is thus *necessarily* very large, and if it is very large, so (by his hypothesis of an invariant relationship) must the other two be. But this explanation obviously comes at a cost: it makes the other two numbers vary over time. Hence the supposed "constants" of physics on which these numbers depend, would also have to vary over time: the Newtonian gravitational "constant", *G*, would have to weaken steadily from much higher values earlier in the universe's history. The consequences for physics generally would thus be radical. Worse, there was no independent evidence for any such variability. Not surprisingly, Dirac's hypothesis did not receive much support, though it did lead Pascual Jordan, for example, to propose a tentative reformulation of the theory of gravity in order to allow for a varying gravitational "constant".

It was Robert Dicke, in a letter to *Nature,* who pointed to an "anthropic" reading of the Dirac hypothesis, one that was intended, effectively, to undermine it.[13] He noted that the third of Dirac's numbers, the age-ratio of the universe, T, depends on the person making the calculation, more specifically on the location that that person has on the universe's timeline. It is thus an *anthropic* value (though Dicke did not employ that term.) Humans could exist, he goes on, only after the universe had produced the heavy elements needed for their evolution. ("It is well known", he drily remarks, "that carbon is required to make physicists").

Further, they could not exist after the Galaxy's stars had died out. Thus, approximate limits are set at both ends of the period at which a physicist could note the supposed "coincidences between T and the force-ratio, and T and the mass-ratio". He then argues that these two "coincidences" could be regarded as significant only if the present value of the Hubble age could have been widely different from what it is. However, it is constrained within the time-line allowing physicists to exist. Furthermore, he shows that if those limits were computed in terms of the known values of the constants involved in them, the value range of the (contingent) age that they define would *have* to be (very roughly) of the order of magnitude of the other two large numbers, the force-ratio and the mass-ratio. It is thus an artifact of the appropriate position on the cosmic timeline of physicists generally. Clearly, the age number is not independent of the other two, as Dirac's puzzle about coincidences takes for granted.

Dicke's emphasis on the selection effect involved in taking the age of the universe as one of the three numbers would, of course, be challenged by Dirac, for whom there was no such effect. If *his* hypothesis were correct, it would not matter at what point in the cosmic time-line the cosmic age were to be computed, whether in the physicist era or outside it. The relation of the age to the other numbers would be the same at all times, involving, of course, a steady change in those numbers. How that change would affect the separate constituents of those numbers would be an open question, and Dicke's computation of the aggregate values of the large numbers themselves would thus become problematic. Dirac's hypothesis would have allowed him a riposte.

However, there is no need to get lost in the teasing tangle of issues involved in the Dirac-Dicke debate.[14] It is aufficient to note that Dicke was underlining a consequence of the Big Bang theory that would shortly come to have a great deal of significance. Since the theory implies that the universe has steadily changed in times past, one must expect selection effects due to the physicist's contingent position on the cosmic timeline. In particular, an observer utilizing astronomical data is viewing the universe, not as it is, but as it was, thus providing a ready way to decide between the two major cosmological models, Big Bang and Steady State. Only the former would show a past distinctly different from the present. This way of deciding is implicitly anthropic in its implications.

Getting "Our" Universe Started

The anthropic flurry so much in evidence in cosmology today began in the early 1970s when quantum theorists began to investigate the novel new field of research opened up by the extension of the postulated Big Bang expansion back to its first moments, when the immense energies would involve all sorts of quantum effects. In 1973, Collins and Hawking came up with a most unexpected, and to them most unwelcome, entailment.[15] The only way, or so they said, for our universe to be as isotropic as it is after so long a period of expansion would be for its initial energy density to be at the borderline between values that would lead to a runaway expansion and values that would lead to rapid collapse. The cosmic geometry would have to be "flat" to begin with to an almost unimaginably precise degree.

This finding was unwelcome because it challenged the Cartesian principle of indifference that had implicitly informed cosmology for so long. Were a "chaos" to be aufficient to give rise to the sort of universe we now have, no question would arise about why its parameters had the evidently contingent initial values they had. But if the present universe severely *constrains* the range of possibilities for a plausible starting point, a question about the significance of that constraint immediately presents itself, or so Collins and Hawking believed. Their guess was that the significance was anthropic in nature and that some sort of selection effect was responsible. The conditions necessary for the development of intelligent life: galaxies, planets, heavy elements, etc., can only occur in a very long-lived universe. However, a long-lived universe is the only kind that we could inhabit. The universe has to be long-lived because we are here!

Unfortunately, a necessary condition does not qualify as an explanation. A further premise is needed, and one immediately suggested itself to the authors. There had already been talk of "branching" worlds in attempts to understand wave-packet collapse in quantum theory. What if there were, not just one world, but a very large number of (actual) worlds characterized by, in effect, the widest range of possible initial conditions, specifically of energy density. In that case, one could expect that there would be one or a small number in that ensemble of worlds satisfying the energy condition that would give rise to our sort of long-lived world, thus permitting the evolution of humans.

The existence of a sufficient number of worlds to make this outcome unsurprising would "explain" why we find ourselves in a world whose initial conditions seem so finely tuned for our eventual appearance. The plurality of worlds possessing the requisite range of initial conditions is the explanation therefore, of what originally seemed puzzling. In another sense, it explains *away* the apparent fine-tuning. There *was* no fine-tuning after all, strictly speaking. Rather, there was a selection effect at work. The kind of universe we find as *we* theorize back to cosmic beginnings is the one that our complex physical make-up effectively "selects" for us.

This pushes the Principle of Indifference up to a much higher level. The unspecified initial condition (the "chaos") is now an ensemble of worlds, not of particles in motion within a single world. To get to "our" sort of world, once more we need not lay any special constraint on what the initial "given" should be like, in terms of energy density at least. The constraint only enters when one asks how many of the worlds within that initial "given" are suitable for the possible evolutionary development of intelligent life.

What is significant here is the constraint that the human presence places on the cosmic starting-point. Exactly the same constraint would be called for by the presence of beetles or oak trees. But there would be no selection effect in their case – they do not pose questions that implicitly assume a selection effect on the part of the questioner. The explanation afforded by postulating a selection over a multiplicity of worlds is thus properly anthropic. But is there actually any kind of anthropic *principle* here?

Carter's "Anthropic Principle"

The anthropic reasoning involved in the Collins-Hawking argument was dignified by Brandon Carter the following year with the label, *anthropic principle:* "What we can expect to observe must be restricted by the conditions necessary for our existence as observers".[16] As it stands, this is, of course, quite trivial-sounding. And Carter complicated matters by introducing also a "strong" version of the principle: "The Universe must be such as to admit the creation of observers within it at some stage". This either reduces to the first ("weak") form or makes an implausible claim, depending on the force given to the term, "must".[17] The ambiguities of the Carter article have given rise to an abundant literature. It may be best to dispense with the notion of a "principle" here; it promises too much. Or if one must have one: "in an expanding universe, watch out for selection effects!"

Finding an initial cosmic parameter constraint where none was expected led physicists to ask: was there any *reason* for it? And a selection effect gives an elegant answer. But, of course, the postulate on which it depends is highly speculative. The vigor with which it was propounded in the 1970s by Hawking, Carter, and many others, testifies as nothing else could to their commitment to the principle of indifference. In the absence of even the slightest independent theoretical reason to suppose that the ensemble of worlds displaying the requisite range of the relevant parameter actually exists, they propounded the many-world scenario as though the solution it offered of the apparent "fine-tuning" puzzle was evidence enough in its support, despite its extraordinary ontological extravagance.

What prompted Collins and Hawking to propose their many-world scenario in the first place was the extreme constraint required for a single parameter, an initial condition: the total energy density. But Carter pointed to another source of constraints of a significantly different sort. The energy density seemed, to all intents and purposes, to be an entirely contingent matter. So far as one could tell, it could have the widest possible finite range that theory allowed over a multiplicity of "worlds" that constituted a single "universe", the universe itself remaining the same kind of universe. But what if the many worlds required by the anthropic solution vary in a more fundamental way, exhibiting different basic constants and different "laws of nature?"

Here came the second great surprise, the second challenge to the cherished principle of indifference. If the strong nuclear force binding nucleons into nuclei were only very slightly weaker relative to the other three fundamental forces, Carter notes, hydrogen would be the only element. If the gravitational force were slightly stronger, planets might not form. If the fine-structure constant were "increased by only a very small amount", likewise planets might not form. So it begins to seem that there is a much more fundamental set of cosmic constraints imposed by the anthropic requirement. Carter was more interested in explaining in this way why the gravitational force should be so extraordinarily weak, relative to the other three forces. (He leads into his article by recalling Dicke's invocation of a selection effect in the context of the large-number coincidences.) However, his implication was clear: one may well find that there are all sorts of *nomic* constraints, constraints on the laws of physics themselves, for a universe to develop in a potentially life-bearing way.

And find such nomic constraints physicists did. Indeed over the next decade it became almost a parlor game to juggle with the balance between the four forces to see what would happen in the ensuing cosmic scenario: only helium if the strong nuclear force were as little as 2% stronger; only hydrogen if it were 5% weaker. The existence of supernovas, of planets, of carbon, was shown to depend fairly sensitively on the values the constants actually have in our universe.[18] An appeal to an anthropic selection effect once again suggested itself.

This time, however, the requisite many-world postulate was much more problematic. It was one thing to conjure up new worlds differing only in their contingent energy densities. It was at least conceivable that an overarching theory might be found in which a single universe might give rise to such a multiplicity of component worlds, causally linked, say, at their origin and hence, in principle at least, theoretically accessible. But *nomic* variation was another matter. Was it even permissible? Might not the relationships between the physical constants be *necessarily* what they are? And what sort of super-theory could be imagined that would govern the sorts of variation that this new appeal to the anthropic entailed?

The Teleological Alternative

The physicists who discovered the different sorts of initial parameter constraint that are implicit in the development of the long-lived, planetbearing, heavy-element-forming universe we know, saw those constraints as significant and dealt with the consequent demand for explanation by invoking the anthropic dimension of the universe. The constraints are a logical consequence of that dimension, if the many-world postulate is granted. But might not the constraints be significant in a quite different sense, one more familiar than a selection effect, when tracing for a reason for something earlier in something later. Might they not be a *teleological* consequence instead?

It was not hard to see how this might be. The familiar story of creation shared by all three Abrahamic faiths postulates a Creator on whom the existence and nature of the world depends.[19] The Biblical story describes a universe whose contents are of the Creator's shaping that is brought into existence at a moment of time. Medieval Christian theologians elaborated on this metaphysical theme and argued, for example, that time itself was the creation of a Being who existed beyond time (Augustine), and that even were the universe to be eternal, as Aristotle had held, it would still need a Creator cause to account for its existence and nature (Aquinas).

But a second premise is also needed, a more specifically theological one. Human beings play a special role in the universe story as this unrolls in the Bible and the Koran. They are said to be made in the Creator's image; they appear to be the object of the Creator's special concern. Their abilities set them apart from the rest of the creation – their combination of reason and free will makes them capable of a response to the Creator (of love, of acknowledgement, of denial) that no other beings can claim. The sacred books of all three faiths tell of a God who is intimately involved in the doings of His people. There is a third premise implicit in all this too; that the Creator is not just an impersonal force or energy of some unimaginable sort but a being capable of concern whose creative agency is guided in some sense by an analogue to what humans would call purpose.

If these three premises are granted, then cosmic constraints are easily explained. To the extent that the advent of the human plays a significant part in the Creator's purposes, it was to be expected that whatever constraints on the original shaping were necessary for that purpose to be fulfilled would in fact have been incorporated by the Creator. If the universe is regarded as a Creation, then the characteristic ways of acting (the "laws of nature") and the initial conditions (e.g. initial energy density) would be set by the Creator's agency. And if the universe is to be "anthropic", capable of evolving to the level of the human, the Creator would simply elect those particular constraints on the limitless possibilities of creation that would make this outcome possible.

This is an anthropic explanation but in yet another sense. It does not appeal directly to human agency as an archaeological explanation might. It does not see all the constraints as a selection effect attributable simply to the anthropic involvement in the posing of the original question. The anthropic dimension appears rather, in the purposes of the Creator – the act of Creation is deemed to have been guided by purpose and the capacity of the universe to be human-bearing was part of that purpose. This conforms to the familiar model of teleological explanation, but of course the context is a far from familiar one, one where the very notion of purpose stretches the limits of analogy.

An important feature of this way of understanding the problematic initial constraints is that, unlike most "Design" arguments of past natural theology, this does not involve any sort of miraculous intervention of the Creator in natural process, no momentary setting aside of the normal operations of the laws of nature. If the notion of a purposive Creator is admitted in the first place, this automatically entails a choice on the Creator's part of the sort of universe that conforms to the Divine purpose. Electing the constraints that in this respect allow the universe to conform is simply implicit in the Creation postulate itself.

In this perspective, the metaphor of "fine-tuning", with its implication of a Tuner, is entirely appropriate. Not surprisingly, critics of the theistic alternative,[20] object to it as implicitly begging the question. I have used the neutral term, "initial parameter constraints", deliberately here and there in order to avoid any misunderstanding. A further reason for this choice is that it focuses attention on the essential element: the constraint on values that according to the principle of indifference one would expect to be unconstrained. The *degree* of constraint makes a difference to the motivation for seeking an explanation – the tighter the constraint, the greater the motivation, as Collins and Hawking recognized.

The teleological alternative is clearly not scientific. It relies on both metaphysical and theological considerations. So those who rule out such considerations and limit explanation in the cosmological domain to natural science alone will, of course, rule out the teleological alternative also. Those who are led to postulate a Creator on traditional metaphysical grounds might still balk at the anthropic move that relies rather more on theological supplementation. But for those who already accept the view of creation implicit in the Jewish, Christian, or Islamic traditions, the teleological alternative could seem a plausible choice. Given their background beliefs, the anthropic solution could neatly account for initial cosmic constraints. The overall explanation is, of course, a theistic one. But it is the tie to the *anthropic* dimension of the universe that makes it work.

Many philosophers of religion were intrigued by these unexpected developments in cosmology, and debate began as to the weight that should be given them. Some took the opportunity to advance a full-scale natural theology centering on cosmology and emphasizing the anthropic-theistic dimension of the new scientific findings.[21] Others were more hesitant – arguments drawn from design that single out features of the world that could best (or only) be explained by invoking the action of a Creator/Designer have not had a good record. Proponents of the new argument responded that it did not, as the earlier arguments had done, rely on an explanatory gap on the side of science – something that science *ought* to be able to explain but apparently cannot. The anthropic argument was in fact prompted in the first instance by scientific findings and continues to rest on those. It proposes an agent-causal explanation of the origin of physical laws and initial cosmic conditions, a topic that lies outside the scope of physical science.

But what if the original *scientific* claim had been in error or at least premature? What if the alleged constraints were, in fact, not necessary? What if the cosmologists could provide not just a possible non-teleological alternative explanation (the anthropic selection effect did not have many defenders at the time), but a reason to believe that there was nothing to explain?

Inflation and After

In 1980, Alan Guth proposed an ingenious, if highly speculative, way of accounting for the present "flatness" of the universe, the problem from which the whole anthropic discussion had begun in 1973, without requiring the almost unimaginably precise "flat" setting of the initial energy density. His proposal affected only the first fraction of a second of the cosmic expansion, leaving the standard Big Bang theory to handle the rest. Suppose a phase transition occurred that brought about a gigantic expansion, multiplying the diameter of the point-like universe by a mind-boggling factor of the order of 10^{50} and then stopping as suddenly as it began. The effect of this would be to force the energy density, no matter what its initial value might have been, towards the desired "flat" value that would then be maintained. No need, then, for any constraint on the initial value – the principle of indifference is once more in command.

Guth's "inflation" model was prompted in the first instance by the apparent parameter constraint that the initial state required. He regarded this latter as a "peculiar situation" that could not be left to stand. The manyworld alternative evidently did not appeal to him. What is interesting is to note how the strong the motivation was on his part and on that of many other leading cosmologists, to get rid of the troublesome initial constraints that gave rise to such strange anthropic fancies. Guth's model, like any first try, faced a number of serious technical difficulties, most of which have been overcome in the years since then by introducing a number of major modifications. Vigorous attempts have been made to find independent observational support for the inflation hypothesis; some success is claimed, for example, in its prediction of the amplitudes of the tiny fluctuations in the cosmic microwave background.[22] Its major promise is to offer hope of explaining how in such an isotropic universe the galaxies could have begun to form – the inflation ratio is so great that quantum fluctuations could have been enormously amplified to form the galactic seeds.

The notion of inflation has itself meanwhile been inflating. Once inflation is admitted, some theorists have argued that it can happen over and over in separate domains, generating new "bubble" universes at no time causally connected with one another. The ensemble of such universes (now "universes", no longer "worlds") has been called a "multiverse". Thus, inflation, originally proposed as a way of *avoiding* the anthropic line of argument has, rather ironically, given fresh life to it. The most speculative of the multiverse theorists, the Russian physicist Andrei Linde, would have every other sort of universe out there somewhere, popping endlessly in and out of existence in an infinite space-time. These universes would exhibit the widest variety of basic constants; their laws, even their space-time dimensionality, would range as far as the theorist's imagination might let them.

It must be emphasized that these last theoretical forays are still not much more than imaginative constructions, a series of ever more daring "what-ifs". But they do bear on the second form of parameter constraint discussed above: the tight constraints on the basic constraints of nature that the development of an anthropic universe calls for. Even if the energy density issue were to be explained away by inflation in our own universe, the more significant parameter constraints are assuredly the nomic ones, those embodied in the laws of physics themselves.

The multiverse scenario would once again restore the principle of indifference, now at an even higher level – that of an ensemble of universes, each displaying a different nomic configuration. The anthropic selection effect would once again come into play: we are in a unique universe among the vast ensemble of "bubble" universes out there, the one that has the "right" mix of physical constants. For the moment, this is no more than the hand-waving of a very bright physicist. Our own universe, even if inflation were added in, still displays the tight nomic constraints that have seemed to many to call out for explanation of some sort. To handle these by invoking a selection effect, a multiverse of the Linde kind would be needed. That price would seem, at the present juncture at least, to be too high. However, its possibility in principle must be kept in mind.

Four Alternatives

In the light of the developments in cosmology of the last thirty years, there would seem to be four, and only four, alternative ways of dealing with the cosmic parameter constraints that have recently come to light. And adjudicating between them is extraordinarily difficult, given the lack of agreement on the proper criteria to employ and given also that cosmology itself is in such a state of explosive development. Still, to finish, it may be worth summing up the prospects for each of the four alternatives.

Happenstance

The constraints on initial cosmic conditions and on the law of physics themselves imposed by the anthropic requirements are taken to be real. But they are declared to be just a matter of chance. The universe just *happens* to be that way: why not? If it were to be at all, it had to be *some* way. And, lucky for us, that is just the way it is.

This is perhaps the least favored of the alternatives. Most scientists are uneasy about this dismissive way of dealing with coincidence, even in a world where quantum chance seems pervasive. Time and again, ever since the original constraint on initial energy density came to light, cosmologists have devised highly speculative theoretical constructions either to show that the supposed parameter constraint is in fact illusory (Guth), or to provide an anthropic means of restoring the valued principle of indifference (Collins and Hawking). They *could* have said: well, that is just the way it was – a matter of chance like the moment at which a radioactive atom disintegrates. But they didn't...

What struck them was, of course, not the initial state alone – considered in isolation, one specification of that state would seem as likely as another. It was the apparently tight correlation between that state and a later history that to all appearances would have been altogether different had the initial state been only *slightly* different on the scale of the parameter itself. Was *this* significant or not? Was there some feature of that later history that might help to explain the correlation? One could argue that this is to push the demand for explanation too far. But physicists are wont to push that demand as far as it will go.

Are the Constraints Real?

A different response would be to question whether the supposed constraints are themselves real. What if they reflect merely a temporary stage in theorydevelopment? If history is any guide, how can one be confident that constraints of this sort will survive theory change? Might there not be a return to the traditionally bland indifference of the initial cosmic state?

In a sense, of course, this has in fact happened, or at the very least a route to it has been opened. The inflation hypothesis was devised in the first place precisely to eliminate the supposed "fine-tuning" as well as to respond to the so-called "horizon" problem. Highly speculative originally, the hypothesis now has attained a degree of respectability. It has had to be extensively reworked along the way, and still faces some difficult challenges. It is not clear that it has entirely eliminated the need for "finetuning" on its own account. But the degree of success that the inflation hypothesis has in fact enjoyed should make one very cautious indeed in claiming the need for parameter constraints in the context of such seemingly contingent parameters as energy density.

The nomic constraints, however, obviously have a quite different status. The basic constraints have been measured with an impressive degree of accuracy. The process by which, say, carbon is formed by helium fusion is reasonably well understood and the consequent parameter constraint is straightforward to formulate. The basic constraints are, of course, theory dependent in numerous ways. But it seems safe to say that later theory, no matter how different it may be, will turn up approximately the same dimensionless numbers. And the numerous constraints that have to be imposed on these numbers if a complex long-lived universe is to be allowed to develop within the framework they define, seem both too specific and too numerous to evaporate entirely, as may well happen with the original "finetuning" claim for energy density. Take a typical example: "A change of only half a percent in the strong force would stop the helium fusion on which the formation of carbon depends..."[23] It might happen, of course, that an alternative account of the formation of carbon will be devised, one that would dispense with the claimed constraint. But then a dozen or more other similar constraints have been pointed out, each dependent on a different theory of how a certain sort of physical transformation occurs. Might they *all* be replaced? Well, perhaps, but it surely seems a very long shot.

There are, however, several other quite different sorts of challenge to the reality of the "fine-tuning" claim. The first is that a later super-theory might well show that these constants, and the laws following from them, are interconnected, that they cannot, in principle, vary independently of one another. So that one could not vary the strong force by 5%, say, without invoking a cascade of other changes. Its present value might, in other words, be dependent on, or even necessitated by, the values of the remaining constants.

Might they not all vary together, then? Would some sort of "finetuning" persist in this case? We simply don't know. It is one thing to suppose that the initial cosmic energy density might have been different; it seems, at least, a contingent parameter. But the matter is otherwise with the laws of physics themselves and with the dimensionless constants that appear in them as pure numbers. Might they have been different? It is not clear that science has an answer to this. What sort of theory would tell us what the universe *might* have been like, even though it isn't that way?

A suggestion that has sometimes been made (by Hawking, among others) is that it may turn out that the present values of the basic constants *have* to be what they are, that a necessity binds the system as a whole. We are surely far from such a theory as matters now stand. One could point to the many unresolved issues in basic physics and cosmology right now: the lack of integration between gravitational and quantum effects, and the growing "darkness" in cosmology, as "dark" energy has been added to "dark" matter as yet another troubling question mark. If the long- sought "final" theory brings all of these elements together, showing that the constants *have* to have the values they do, that no other set is possible, it would raise a host of questions as to what that would even mean. And, of course, even though such a theory would explain, in a sense, why the constants have the values they have, it would leave open the further question: is this still not itself significant? Might not a *further* question be in order here?

A very different sort of objection to the whole idea of "fine-tuning" or significant cosmic parameter constraint is that for it to have any force, there would have to be a way to define the values of the probabilities involved. This is impossible, it is urged, since for one thing the required measure function is non-normalizable. So the intuitions to which physicists like Hawking and philosophers like Leslie appeal in this context are no better than that – intuitions, and intuitions with nothing to back them.[24]

The issues here are complicated and I cannot do justice to them in short space. First, the critics are right in one regard. I do not think that numerical probabilities *can* be assigned here, even though physicists as eminent as Hawking and more recently Lee Smolin[25] have not hesitated to do so. It is important to distinguish here between the two types of possible constraint: on a parameter like initial energy density which could take any one of a wide range of values as far as the relevant theories are concerned, or on a basic physical constant like the strength-ratio between one of the four fundamental forces and the other three. The notion of fine-tuning was originally attached to the first of these only. In this case, the idea of a small constrained range compared to a much larger possible one seems to offer a handhold for a probability estimate.

Some of the critics at this point assume that the larger range here would have to be infinite, and then have little difficulty in showing that this would lead to a zero probability for *any* finite range of the parameter.[26] Of course none of the physicists involved want to allow the possible values of the energy density to range to infinity. Their assumption is that the range is very large, so far as the relevant theories are concerned, large enough to make the postulated constraint significant. This would not yield a numerical estimate for the probability. The further implicit assumption that each interval of the possible range is equally probable (required by normalizability in the absence of the alternative, a density function over the range) could also be problematic. Far more problematic, however, would be the application of probabilistic notions to possible variations in the fundamental physical constants. It is hard to see how this could be theoretically grounded, since we obviously have no theory as to how these constants *might* vary. So that defenders of the necessity for "fine-tuning" would seem to be ill-advised to introduce numerical probability estimates of any sort here.

But who says that in the absence of a numerical estimate that Bayesians can work with, the notion of a required parameter constraint collapses? Such a constraint violates the Principle of Indifference and the tighter the constraint that theory ordains relative to the plausible finite range of the parameter, the more pressing the question becomes: can there be an explanation for the fact that the initial cosmic state *was so* constrained relative to some later feature of the universe? No probability estimates are needed to make this question meaningful. It is enough that a variation of only a tiny percentage in the value of a dimensionless constant would exclude entirely some later cosmic development like the formation of helium. What fine-tuning amounts to in cases like this has nothing to do with probability estimates.

As we have seen, there are serious questions about the direction future theory may take and about how to understand the idea of nomic variability that underlies the supposition that the fundamental constants might have had values other than those they actually do have. On the assumption, however, that the "fine-tuning" is real, two alternatives remain, both of them anthropic.

Many-World

Given a multiplicity of actual worlds over which the relevant parameter ranges sufficiently widely, an anthropic selection effect would, as we have seen, explain the fact that the parameter took on the constrained value it did. It would explain away the appearance of a deliberate fine-tuning on the part of a Maker. But there would have to be reason other than this alone to make such an ontologically extravagant postulate plausible. That is, there would have to be a physical theory implying the existence of this profusion of worlds, a theory for which there is independent evidence in the world we inhabit. If fine-tuning of the second kind, that is, of the basic constants, is in question, than one would need a theory requiring such a multiverse.

There are such theories, notable among them those of Andrei Linde. Their existence should at least make us pause. But whether sheer technical ingenuity in the absence of testable consequence is enough to confer plausibility is questionable. *If* a multiverse of the Linde type, displaying physical laws of all sorts in different bubbles of space-time, were to exist, an anthropic selection effect would assuredly operate. However, this of itself is far from enough to warrant belief that we do inhabit such a bubble. Much more would be needed to make the multiverse alternative an appealing option.

Purposive Fine-tuning

The remaining alternative, assuming the initial cosmic constraints are real, is to suppose that they have been purposely "tuned" in this way in order to accomplish the ends of the Tuner, and these are further assumed to include the existence of humans. (Obviously a non-anthropic motive could also be postulated – there are many other possible reasons why a Tuner might want a universe of the sort that would also accommodate humans).

Appeal to a *Creator* has a particular advantage over the multiverse alternative in the context of properly *nomic* constraints. As we have seen, when we ask how the basic constants might in principle vary and what limits one should impose on such a variation given the physics of the universe (our only source of scientific evidence), we run into all sorts of difficulty. There may be relations of necessity between the constants of which we remain unaware, though it would be very difficult to believe that no other physics than one featuring the present values of the basic constants is even metaphysically possible. But even if there are internal limits of necessity on these constants in the context of the universe we have, we could ask: why a universe of *this* sort – one which makes the advent of the human possible – given that an entirely different universe, one governed by a quite different physics, seems metaphysically possible. Appeal to a *Creator* (not to a Maker constrained by the materials available) would provide an answer to this.

This is not a scientific hypothesis. Recalling the archaeologist who finds an incised bone deep in the earth that may or may not testify to a human presence in that location at a specific period, one would have to investigate the likelihood of such a Creator in the first place. This would necessarily lead, as we have seen, in metaphysical and theological directions. There is no way that the natural sciences, on their own resources, could support, or for that matter exclude, such a hypothesis. Does this rule it out, as a good many of those who have entered the anthropic debate clearly assume? I would say not, but to hold this possibility open would, of course, need extended argument on its own account.

A Creator of the sort that informs the Western religious tradition would, as we have seen, explain fine-tuning, to the extent that such fine-tuning exists. Such a Creator according to this tradition is already held to be responsible for the existence and nature of the cosmos, and would therefore incorporate in the work of Creation whatever further constraints the Creator's purposes might include. This is not a challenge to the sciences; it supplements their findings by covering an issue of cosmic origins that can reasonably be said to lie outside their domain. (Again, argument is needed here in regard to recent claims about vacuum fluctuations and the like.) The crucial question here is the ancient metaphysical one: should we simply take the existence of the universe for granted, on the assumption that causal argument of the conventional sort fails at that level, or should we postulate a transcendent causal agency of some kind, one with an interest in human existence besides, for the anthropic argument to work?[27]

Suppose one were to pass over the difficulties in establishing the reality of the fine-tuning feature of cosmic origins itself, would the anthropic argument offer *independent* reason to believe in the existence of a Maker of the sort the argument would require? I am hesitant to answer this in the affirmative, even though the logic of theory-support might suggest that I should.[28] And the reason is the same as that already outlined in the case of the other anthropic argument, the many-world one. In both cases, a giant ontological leap is required. To confer some initial plausibility and to allay the suspicion that the proposed explanation is *ad hoc,* entirely contrived, one has to show that the hypothesis is coherent, that the entity being proposed by way of explanation is accessible to our theorizing, scientific in one case, metaphysical and theological in the other. With this assurance given, the anthropic explanation might be held to confer additional plausibility. However, then we run into the other difficulty: how sure can we be of the explanation itself – the alleged *fact* of fine-tuning? All in all, then, the anthropic argument would seem a weak reed as a motive for belief in a Creator. It is *consonant* with such belief if that belief is already there. That in itself is enough, certainly, to commend it to our interest.

In the end, however, all four alternatives have to be kept in mind. Balancing their relative likelihood, even in the roughest way, makes demands of an unusual epistemic sort. That is what has made this issue at once so fascinating and so incapable of agreed solution.

NOTES

[1] Ernan McMullin, "Indifference Principle and Anthropic Principle in Cosmology", *Studies in the History and Philosophy of Science*, 24, 1993, pp. 359-389.

[2] Isaac Newton, *Principia,* Motte-Cajori translation (Berkeley: University of California Press, 1966), p. 544.

[3] Newton's first letter to Bentley, in I. B. Cohen, ea., *Isaac Newton's Papers and Letters on Natural Philosophy* (Cambridge, Mass.: Harvard University Press, 1958), p. 282.

[4] Richard Bentley, "A Confutation of Atheism from the Origin and Frame of the World", in Cohen, ea., *Papers and Letters,* 1693, p. 363.

[5] See Sherrilyn Roush, "Copernicus, Kant, and the Anthropic Cosmological Principles", *Studies in the History and Philosophy of Modern Physics,* 2002, in press.

[6] Immanuel Kant, *Critique of Pure Reason,* A42, B59, transl. N. K. Smith (New York: Macmillan, 1965), p. 82.

[7] Arthur S. Eddington, *New Pathways in Science* (Cambridge: Cambridge University Press, 1935) Chapter 11, "The Constants of Nature"; *The Philosophy of Physical Science* (Cambridge: Cambridge University Press, 1939) Chapter 4, "The Scope of Epistemological Method". In his 1935 book, he identifies the fourth number with the ratio of the radius of curvature of space-time to the mean between the proton and electron wavelengths in wave mechanics; it is (he says) about 10^{39}. In the 1939 book, the fourth number, now called the "cosmical number", is identified with the number of particles (protons and electrons) in the universe, arrived at as the number of particles sufficient to cause the cosmic space to close. See also the Appendix to his posthumous work, *Fundamental Theory* (Cambridge: Cambridge University Press, 1946) "The Evaluation of the Cosmical Constant".

[8] *The Philosophy of Physical Science*, p. 60.

[9] *Fundamental Theory*, p. 265.

[10] *The Philosophy of Physical Science*, p. 62.

[11] See especially Susan Stebbing, *Philosophy and the Physicists* (London: Methuen, 1937).

[12] See the abundant documentation of the continuing discussion in John D. Barrow and Frank J. Tipler, *The Anthropic Cosmological Principle* (New York: Oxford University Press, 1986), notes to Chapter 4.

[13] Robert Dicke, "Dirac's Cosmology and Mach's Principle", *Nature,* 192, 1961, pp. 440-1.

[14] The strength of the original Dirac hypothesis lies not so much in the alleged coincidence between the three large numbers (which in the end he left unexplained) as in the fact that by making the relation between them invariant and hence time dependent, he was able to suggest why they should be so large. The Dicke alternative denies him that invariance (the relation holds only within the "physicist" era). But then Dicke has (in Dirac's eyes) to explain the large size of the numbers and why the values of the force-ratio and the mass-ratio are approximately the same. This Dicke tried to do by invoking the Mach Principle, along lines that Dirac might well have questioned.

[15] C. B. Collins and S. W. Hawking, "Why is the Universe Isotropic?" *Astrophysical Journal,* 180, 1973, pp. 317-334.

[16] Brandon Carter, "Large Number Coincidences and the Anthropic Principle in Cosmology" (1973), reprinted in J. Leslie, ea., *Physical Cosmology and Philosophy* (New York: Macmillan, 1990), pp. 125133.

[17] E. McMullin, "Indifference Principle and Anthropic Principle in Cosmology", *Studies in the History and Philosophy of Science,* 24, 1993, pp. 372-377.

[18] For a review, see Barrow and Tipler, *The Anthropic Cosmological Principle,* still by far the best source for the technical minutiae of the anthropic arguments. See also John Leslie, *Universes* (London: Routledge, 1989).

[19] See E. McMullin, "How Should Cosmology Relate to Theology?" in A. R. Peacocke, ea., *The Sciences and Theology in the Twentieth Century* (Notre Dame: University of Notre Dame Press, 1981), pp. 1757, at pp. 40-52.

[20] Notably Adolf Grunbaum, in a series of publications. See, for example, "The Pseduo-problem of Creation in Physical Cosmology", reprinted in Leslie, ea., *Physical Cosmology and Philosophy,* pp. 92-112.

[21] See, for example, M. A. Corey, *God and the New Cosmology: The Anthropic Design Argument* (Lanham, Maryland: Rowman and Littlefield, 1993).

[22] For a recent and thorough survey of the present state of the evidence regarding the inflation hypothesis, see A. Liddle and D. Lyth, *Cosmological Inflation and Large-Scale Structure* (Cambridge: Cambridge University Press, 2000).

[23] To quote a recent estimate by Austrian physicist Heinz Oberhummer, reported in the *New York Times,* October 29, 2002.

[24] See, for example, Neil A. Manson, "There is No Adequate Definition of 'Fine-tuned for Life'", *Inquiry,* 43, 2000, pp. 341-352; Timothy McGrew, Lydia McGrew and Eric Vestrup, "Probabilities and the Finetuning Argument: A Skeptical View", *Mind,* 110, 2001, pp. 1027-1037.

[25] Lee Smolin, *The Life of the Cosmos* (New York: Oxford University Press, 1997).

[26] Manson infers that "McMullin's indifference principle" leads to this absurd conclusion, one that would make the probability of a lifepermitting universe zero ("No adequate definition…", pp. 346-7). There is a multiple misunderstanding here. First, it is "my" principle, only in the sense that I defined and named it as an implicit regulative principle in modern cosmology (McMullin, "Indifference principle…"). I do not myself defend it as a constitutive principle. Second, the indifference principle does not (as Manson supposes) imply that the energy density could take an infinity of possible values. All that the principle does is *exclude* the possibility that a constraint would have to be set on an initial parameter, like energy density, in order to arrive at the sort of universe we now live in. This in no way implies that the parameter could take on any of an infinite number of possible values.

[27] Wes Morriston defends a negative answer to this last query in his "Must the Beginning of the Universe have a Personal Cause?" *Faith and Philosophy,* 17, 2000, pp. 149-169. In a rejoinder William Craig proposes a positive one: *Faith and Philosophy,* 19, 2002, pp. 94-105.

[28] As some critics have urged. For an early defense of the view presented here, see "How Should Cosmology Relate to Theology?" in A. Peacocke, ea., *The Sciences and Theology in the Twentieth Century* (Notre Dame: University of Notre Dame Press, 1981), pp. 17-57.

International Symposium on Astrophysics Research and on the Dialogue between Science and Religion, Vatican Observatory, 2003
C. D. Impey & C. E. Petry, editors

Scientific Cosmology: A New Challenge to Theology

Nancey Claire Murphy

Fuller Theological Seminary, Pasadena

Introduction

The purpose of this paper is to show the relevance of recent developments in astronomy and scientific cosmology to theology. I have to say that I'll be focusing on Christian theology due to my lack of expertise in other religious traditions. However, what I say will often be relevant to Jews and Muslims as well, especially my historical remarks, because in the Middle Ages Jews, Muslims, and Christians were all busy borrowing ideas from one another. I shall not be speaking of Christian theology in general, but will concentrate on one aspect of Christian teaching, the doctrine of creation.

Here is my thesis: in the Middle Ages there was a consensus among theologians that the doctrine of God's creation of the universe was relevant to all sorts of cosmological issues, such as the nature of time and the question of whether the universe had a beginning. However, due to a variety of factors in the modern period, most theologians concluded that theology in general and the doctrine of creation in particular are *irrelevant* to the big cosmological questions – theology is basically about *humankind's* relation to God. The ironic development in our own day is that science is now putting all of those big cosmological questions back on the table. And yet it is an important challenge to current theologians to deal with them.

The issues in current scientific cosmology that I shall deal with include the following: first, assorted attempts to provide a scientific explanation of the cause of the Big Bang; second, the "anthropic" implications of the apparent fine-tuning of the cosmological constants; third, the nature of the laws of nature themselves; fourth, the likelihood of life elsewhere in the universe; and, finally, the status of theories about the end of the universe.

Before turning to science, let me list the points upon which theologians agreed, from the early centuries of the Christian era through the Protestant Reformation in the sixteenth century. These are all issues relating to Christian teaching about creation. First, the universe was created by God out of nothing – the common terminology is the Latin: creation *ex nihilo.* Second is the conviction that God alone can create. Third, God created freely for the sake of love. Fourth, creation involves temporal origin; and fifth, the universe is essentially good. This very brief summary covers the tensions among these convictions, as well as the differences in details of interpretation from one theologian to another.

Nonetheless, this consensus provided foundational components of the dominant world view in Western culture until the Enlightenment. However, there are a variety of modern developments, both philosophical and scientific, that have led not only to the rejection of these assumptions in secular culture, but have so eroded the doctrine's place in theology that, as one American theologian says, it does not now have a vivid and compelling life in the churches.[1]

The historical factors leading to the loss of consensus on the meaning of creation are many. Jesuit theologian and historian Michael Buckley has argued persuasively that the way was prepared for this breakdown by pre-Enlightenment theologies in which attempts were made to support Christian faith by means of philosophy alone, apart from any theological reasoning. Such arguments could support at best *thin* doctrines of God and creation.[2]

Other factors in the demise of the traditional consensus related in one way or another to the rise of modern science and, in particular, to scientific accounts of the origins of life and of the universe as a whole. I suggest that there have been four sorts of response to these scientific developments. One pair of responses takes scientific accounts of origins to be in conflict with the doctrine of creation, and rejects one or the other – either the science or the very idea of creation.[3] The rejection of science is still an all-toocommon response among conservative Protestants in the United States.[4] A more interesting sort of response was typified by Isaac Newton's construction of a physico-theology in which God's creative and providential role was adapted to the needs of his physics and, in exchange, God's existence was assured by that very function in the system. Accounts of God's relation to nature still tend to suffer from the deism for which Newton's strategy paved the way. Deism is the technical term for the view that God created the universe in the beginning and then left it to run on its own.

The response to modern science that has had the most pervasive effect on mainline theology as a whole, and the doctrine of creation in particular, was the attempt by nineteenth-century liberal theologians to *immunize* theology from science. These strategies were inspired by the philosopher Immanuel Kant, who made a sharp distinction between the phenomenal world, known to science, and the noumenal world, known by means of moral intuition. God was to be associated only with the noumenal world.

Friedrich Schleiermacher, often called the father of modern theology, relocated God and religion within the aesthetic sphere – the world of human feeling. Thus occurred the so-called Copernican revolution in theology, the "turn to the subject". For Schleiermacher, Christian doctrines are "accounts of the Christian religious affections set forth in speech".[5] So the doctrine of creation is an expression of the Christian's *felt awareness* of the absolute dependence of all things on God, and has *nothing* to do with the questions of origins.

In a rich variety of ways, mainline theologies up to the present have followed Schleiermacher in this anthropocentrizing tendency. For example, there is the contemporary American theologian Langdon Gilkey's concentration on the "religious meaning" of doctrines, which he defines as the attitude toward reality, life, and its meaning that the symbol (i.e., doctrine) expresses. In the case of the doctrine of creation this is "an attitude toward God, the world, and human life in space and time".[6]

While this turn to the subject began in Protestant theology, is has thoroughly affected (or, as I would prefer to say, infected) Catholic theology as well. For instance, in Karl Rahner's *Foundations of Christian Faith* the word "creation" does not occur in the table of contents; instead there is an entry on "Man's Relation to his Transcendent Ground: Creatureliness". Here the focus has shifted from *God's* act as creator to *human* experience of relationship to God.[7]

It is now widely recognized that this anthropocentrizing of the doctrine of creation has had (at least) two negative effects. First, as already noted, it has resulted in the marginalization of the doctrine both in systematic theology and in the life of the church. Second, it has left many contemporary Christians without adequate guidance for relating to nature.[8]

Much popular writing by scientists and philosophers today might be described, borrowing Gilkey's term, as reflections on the "religious meaning" of current debates in scientific cosmology. The thesis of this paper is that, in light of the importance of these discussions in contemporary culture, theologians can no longer reduce the doctrine of creation to a reflection on humankind's relation to God, but must return to a consideration of the questions of origins, temporality, finitude, and others that were once thought of as central to the meaning of the doctrine. I shall claim that while science does not always support traditional Christian convictions, it certainly shows that most of the issues comprising the earlier theological consensus are back on the table.

My plan in the next five sections of this paper is to address briefly some of the issues pertinent to the older theological consensus on creation and then to show how contemporary developments in cosmology have led to the revival of those discussions.

Creation in Time

The most obvious of the theological issues that has been reopened by developments in cosmology is the question of whether God created the universe in time or from eternity. That is, did the universe have a temporal beginning, or has it always existed, with its existence dependent upon God? Recall from what I said earlier that by the time Schleiermacher published his systematic theology (this was in the 1830s) it was being argued that whether or not the universe had a beginning in time was irrelevant to the doctrine of creation. It had been quite otherwise in earlier Christian thought. For example, in confronting Aristotle's argument for the eternity of the universe, medieval theologian Thomas Aquinas claimed that Scripture teaches that the universe was in fact created in time, but he argued that the notion of an eternal universe is *not* incompatible with the doctrine of creation. This is because the doctrine is essentially about the contingency of the universe, that is, its dependence for its existence on the will of God. Thus, God could have created from all eternity.[9] In contrast, Thomas s contemporary Bonaventure argued that the eternity of the universe was *inconceivable* in that it is impossible to add to an infinite number or to pass through an infinite series. Thus, if time were eternal the world would never have arrived at the present day; yet, it is clear that it has.[10]

Copernicus overturned the Ptolemaic conceptions of the organization and of the motion of the universe but not the conception of the universe as eternal and static. However, the initial development of the science of thermodynamics in the mid-nineteenth century presented problems for the assumption of an *eternal* universe – problems that Bonaventure would have appreciated. If physical systems can undergo irreversible change at a finite rate, then they will have completed those changes an infinite time ago and we could not be observing any such changes today.[11]

As is well known, in the 1920s astronomers had to give up the idea that the universe is *static*. Its observable expansion forms the basis for the Big Bang theory of the origin of the universe. This sudden origin was immediately interpreted by some – believers and atheists alike – as confirmation of the traditional account of creation as temporal origination. Cooler heads refused the temptation to claim that science had shown the truth of the doctrine; Ernan McMullin's account of the science and theology as consonant has been judged by many to be the most reasonable. He says, "if the universe began in time through the act of a Creator, from our vantage point it would look something like the Big Bang that cosmologists are now talking about".[12]

More recent (and still highly speculative) developments in theories of origins threaten this tidy consonance. A variety of cosmologists have attempted to go beyond what was once thought to be the absolute explanatory limit of science and explain the origin of the Big Bang itself.[13] One of these is Andrei Linde, who speculates that our universe started out as a very small bubble in space-time; the bubble s swift inflation produced the Big Bang.[14] While the notion that the universe must have expanded at a "fantastic" rate at the beginning is widely accepted, Linde's assumption that our universe is but one small bubble in an infinite assemblage of universes I take to be highly controversial. For our purposes, though, it is interesting to see how the very possibility raises again the centuries-old questions of whether the universe is eternal and whether an infinite universe is even conceivable – and in this case, whether an infinite series of universes is conceivable.

So it is clear that if Linde's cosmology were to become widely accepted it would occasion major rethinking of Christian assumptions. One value I see in his account is as follows: the monotheistic faiths are contrasted with Eastern religions in that the latter assume cyclical recreations of the universe. It has long been argued that the linear view of history associated with the monotheistic religions is required to insure the meaningfulness of history and of human endeavor. Yet it seems plausible to ask why an eternal God would create only one comparatively short-lived universe. The so called "principle of plenitude", used by Augustine and others to account for the variety of forms of being *in* the universe, could easily be extended, it seems to me, to allow for the expectation that a God who creates as many forms of being as possible would also create as many universes as possible. Yet, such a plenitude of universes would not count against the meaningfulness of history since each universe would be self-contained and unique.

Creation out of Nothing, Contingency and Finitude

In this section I first turn to another development in scientific cosmology – the "quantum cosmology" developed primarily by Stephen Hawking. We shall see that this theory, whether or not it is ever confirmed scientifically, reopens several of the traditional theological debates.

Hawking's work (as I understand it) depends on recognition that very early in the history of the universe there was a time when the universe would have been compressed enough in size for quantum effects to be significant. Because quantum events occur without causes in the classical sense, this raises the question whether the origin of the universe can be explained without cause, that is, as the result of quantum fluctuations.

In addition, at this scale, the fluctuations would affect space-time itself. Hawking has argued that before 10^{-43} seconds into the universe's existence, space and time would not have been distinguishable as they are now.[15] Paul Davies says, "one might say that time emerges gradually from space", so there is "no actual 'first moment' of time, no absolute beginning at a singular origin". Nevertheless, this does not mean that the universe is infinitely old; time is limited in the past but has no boundary.[16]

A number of authors have commented on possible theological implications of Hawking's work, including Hawking himself. "So long as the universe had a beginning", he says, "we could suppose it had a creator. But if the universe is completely self-contained, having no boundary or edge, it would have neither beginning nor end: it would simply be".[17] Hawking is clearly mistaken in believing that the absence of a temporal starting point eliminates any necessity for a creator since, as we have just seen, traditional accounts such as Thomas's focus on the contingency of the universe – the doctrine of creation is *essentially* an answer to the question of why there is a universe at all, and only *accidentally* addresses the issue of its temporal origin.

However, Hawking's cosmology does legitimately raise a set of related theological issues. The first of these is the traditional emphasis that God's creation is *ex nihilo,* out of nothing. This notion was developed in the second century as the Christian response to a variety of Greek cosmologies. It is based on Old Testament texts that are now seen to reflect explicit rejection of Babylonian creation myths, according to which the world is made of the severed body of a slain goddess.[18] The doctrine of creation *ex nihilo* serves a variety of purposes in Christian theology: it maintains God's transcendence, over and against views that the universe is somehow a part of or an emanation from God. It maintains God's sovereignty, over and against the view that God's creative activity was constrained by the limitations of pre-existing matter, and thus provides grounds for the goodness of the created world, over and against views based on the essential evil of matter itself. Finally, it emphasizes the contingency of the universe – it could have been the case that there was nothing besides God.

Hawking's proposal does not provide a genuine analogy to the universe's origination out of nothing, since the coming into being of the universe, on his theory, presupposes the existence of a quantum vacuum, as well as the laws of quantum physics.[19] It does provoke some thought about what the word "nothing" really means, and more importantly it provokes thought about the nature and status of the laws of nature themselves, a topic to be addressed shortly. Hawking's work (along with that of many others) does confirm Augustine's insight that space and time must themselves be a part of the created order.

Wim Drees has argued that Hawking's elimination of an initial event with a special status recalls Christians to a traditional emphasis on God's creative activity understood as sustaining the universe in existence. All moments in Hawking's theory have a similar relation to the Creator. "It is a nice feature of this quantum cosmology", says Drees, "that that part of the content of creation *ex nihilo* which was supposed to be the most decoupled from science, namely the 'sustaining', can be seen as the most natural part in the context of the theory".[20]

The most subtle and comprehensive response to quantum cosmology, I believe, is that of Robert J. Russell. He points out that Hawking's theory requires theologians to make conceptual distinctions that have been passed over in earlier discussions. First, the concept of a temporally finite creation is distinguishable from the claim that there was a first event, designated as occurring at time zero ($t = 0$). Second, it forces theologians to grapple with the very concept(s) of time that they presuppose.[21] A further conceptual distinction pressed upon us by Hawking's work is a distinction between finitude and boundedness, since Hawking's universe is finite but temporally unbounded.[22] Thus, theologians can no longer make a simple distinction, as Thomas and Bonaventure were able to do, between a finite universe created "in time" and an infinite universe created "from eternity".

The Goodness of Creation and the Problem of Evil

Two related aspects of the traditional understanding of God's creative activity support the conviction that the created world is essentially good. As already mentioned, part of the reason for saying that the universe is created out of nothing is to deny the possibility of intrinsic evil in the universe due to the evil or limitation of any pre-existing materials from which God was constrained to create. Second, God's creation was not *necessitated* by anything within or outside of God. Rather, creation was *motivated* by God's goodness, by love. In this way, Thomas argued that God created in order to diffuse his goodness.[23]

It is against this theological background that the problem of evil appears. So when theologians address the problem of evil they have traditionally done so in connection with the doctrine of creation. The problem of evil arises from the fact that, if God is all good and all powerful, one would expect God to eliminate evil from the world, yet clearly God has not done so. It has become common to distinguish among three kinds of evil, designated as moral, natural, and metaphysical evil. The difficult theological task has been to reconcile these with the assumption of the ultimate (or original) goodness of the created order. Moral evil – that is, human sin – has always been the easiest to account for. A pressing current problem, though, is to reconcile the supposed free will of human sinners with deterministic natural laws, but that is a large enough topic for another conference. I shall focus here on natural evil – that is, the (apparent) disorder in nature and the suffering it causes for humans and animals – and on the closely related topic of limitation, often called metaphysical evil.

Augustine, the great fifth century synthesizer of Christian theology, produced an elaborate set of answers to these three interrelated problems, in all cases drawing heavily on the doctrine of creation *ex nihilo*. Unfortunately, his answer to the problem of natural evil was also dependent on an account of the human Fall as an historical event and especially on the notion of the fall of the angels; both premises are highly questionable on biblical grounds as well as scientifically.

A first step in providing a credible treatment of natural evil is to recognize that all purely natural evil is a simple consequence of the regular working of the laws of nature. Children fall and injure themselves because of the law of gravity; mountain-climbers freeze and people starve to death because of the laws of thermodynamics; deadly bacteria and viruses evolve by means of the same biological laws that have produced ourselves. Of course, much suffering, especially human suffering, is a result of both natural and moral evil – for example, famines are often produced by a combination of factors such as drought and war.

The early modern philosopher G.W.F. Leibniz was much maligned for his claim that this is "the best of all possible worlds". He argued that the more we understand the interconnectedness among things and events the less we can imagine any world that preserves all the goods of this one and eliminates all of the evils. This observation can be all the better supported in light of current science, especially by noting the *connections* that can be drawn among the laws of the various sciences, from physics to sociology. One important contributor to the goodness of a world, in Leibniz's view, is the feature whereby the most results are produced in the simplest ways. On this account, Leibniz would have considered his thesis about the best of all possible worlds to be confirmed by the unification of the laws of physics that we have witnessed in the past decades, and especially by current hopes of finding a theory of everything, a single theory that governs all four of the basic physical forces.

Even more pertinent to Leibniz's argument is the current discussion of fine-tuning and the anthropic principle. Here I refer to the work of scientists such as John Barrow and Frank Tipler, who have marshalled calculations showing that the evolution of life in the universe is dependent upon exquisitely fine balances among the forces and quantities of basic physics.[24] These calculations have led to reflections on the abstract possibility of a vast number of different sets of physical laws. These reflections, in turn, allow us to make better sense of Leibniz's notion of God selecting the best of all possible worlds. Here God must select, from among a number of possible worlds, one of the incredibly small number in which the development of life would be possible.

"Metaphysical evil", in the tradition, refers to the basic fact of finitude and limitation within the universe.[25] Metaphysical evil has regularly been seen as either a condition or an occasion for both natural and moral evil. I believe light can be shed on this ancient idea by focusing on one particular law of nature – the second law of thermodynamics.

Robert Russell argues that entropy is "a prefiguring of evil on the physical level. ...Evil is likened to a disorder, a disfunction in an organism, an obstruction to growth or an imperfection in being. Entropy refers to such disorder, measuring the dissipation of a system, the fracturing of a whole...We need only to think of the pain and cost of natural disasters... to recognize the extent of suffering in this world. All these are rooted in the press of entropy, the relentless disintegration of form, environment, organism".[26] Furthermore, the constant need to replenish the human body – the need for food and other forms of energy – is the cause of much moral evil. While we may dream of a world without this constant loss and degradation of energy, Russell argues on the basis of the work of Ilya Prigogine that the second law of thermodynamics actually plays a necessary catalytic role in the development of higher forms of order, particularly in the development of life.

So we can see that these scientific considerations give theologians additional resources for dealing with evil. Thomas claimed that God caused natural evil *per accidens* – by accident. I prefer to make the same point by saying that natural and metaphysical evil are both necessary but unwanted *by-products* of conditions that were required to furfill God's purposes – including particularly the existence of beings who could freely return his love.[27] One obvious condition for such beings is an orderly, lawlike universe. And now we can see, in addition, that human life could *only* exist in a universe that operates according to laws practically indistinguishable from those we observe to obtain. If the existence of intelligent life is central to God's purposes in creating a universe, then this universe is *one of* the best of all possible worlds. Roger Penrose puts it this way: in order to produce a universe resembling the one in which we live the Creator would have to aim for an absurdly tiny volume of the phase space of possible universes.[28]

Continuing Creation and the Laws of Nature

The emphasis in the Christian tradition on creation out of nothing has tended to obscure the fact that while the Bible does describe God as calling some things into being by his word, it also depicts God as crafting new beings out of preexisting materials – humans from earth, for instance. Thus, *creatio ex nihilo* needs to be complemented by a doctrine of God's continuing creation. This doctrine, as well as traditional accounts of special providence and miracles, has been difficult to sustain since the rise of modern science. The problem comes largely from the concept of the laws of nature: if all natural events are the product of the operation of natural laws, then where is there room for God's action, if not to override those very laws that God has ordained? This is an ironic development, since the concept of the laws of nature was originally used to give an account of God's *ubiquitous* action in nature.

Here is another area where developments in cosmology have put onto the table for discussion a concept with which theologians must reckon. It may be the most difficult obstacle to giving a theologically adequate account of God's relation to nature.

Paul Davies has devoted an entire chapter in his book, *The Mind of God* to questions about the status of the laws. He first surveys the history of the idea and points out that the view of laws as imposed upon matter, rather than inherent in it, was originally a medieval theological innovation meant to defend God's sovereignty over creatures. Early modern scientists believed that by discovering these laws they were uncovering God's rational design for the universe. As long as the laws were thought to be rooted in God their existence was no more remarkable than that of matter. But with the divine underpinning removed, says Davies, "their existence becomes a profound mystery. Where do they come from? ...Are the laws simply *there* – freefloating, so to speak – or should we abandon the very notion of laws of nature as an unnecessary hangover from a religious past?"[29]

Davies' own conclusion is that the laws must be real in somewhat the same way as Plato's Ideas. In fact he attributes strangely God-like features to them: they are universal, absolute, eternal, and omnipotent. However, I believe that there is an insuperable problem for this Platonic account of the laws of nature; it is analogous to the problem of explaining how Plato's transcendent Ideas actually form material beings. Here the problem is *how* such laws "govern" matter. The analogy with laws governing human behavior, upon which the concept was originally based, breaks down when we have inert matter in place of conscious and cooperative agents.

The issues regarding the laws of nature become more complex in discussions of fine-tuning and multiple universes. Some theorists speculate that there is only one logically consistent set of laws. Note that if this is the case then it raises the question of God's freedom in creating. God would still be free to create or not, but not free to choose among possible worlds. Others assume that there are vastly many possible sets of physical laws. If so, are all of those sets of laws "there" from eternity? What causes one or more sets to become instantiated? Or do the laws only come into existence with the world they govern? If so, what determines the character of that particular world?[30]

I am suspicious that, outside of the theistic worldview in which it was born, the very idea of the laws of nature is incoherent. My own speculations in this area have led me to a position closer to Aristotle's and Thomas's than to Newton's: entities in the world behave the way they do because of innate powers and tendencies. The regularities that we observe are a result of these intrinsic tendencies, not of transcendent laws. I have argued that just such a radical reconception of the nature of matter and of causation is needed in order to solve the problem of divine action.[31] This is clearly a topic that needs much further consideration, and, as I mentioned above, may be the most important issue to explore in order to understand God's relation to nature.

Hierarchy, Teleology and the Place of Humankind in Creation

In a previous section I mentioned the place that the creation of humankind has had in the traditional account of creation. This emphasis clearly derives from the creation stories in the first book of the Bible, but it also follows from teachings about God's motive in creating – namely, love. God loves all creatures, but aims particularly to create beings who can consciously receive and freely return that love.

The traditional account of creation has also maintained that the wisdom and power of God are manifested in the perfect organization of the whole of the created order. Ancient and medieval theologians accepted the widely held notion of the hierarchy of being, ranging from inanimate material objects through the various levels of living beings to the spiritual. Every kind of being was seen to contribute to the harmony of the whole. Combining this notion with the thesis of the centrality of humankind in God's ordering of creation, it was all too easy to draw the conclusion that the lower orders of being exist primarily for the good of humans.[32]

A number of scientific developments have raised the question of intelligent life elsewhere in the universe. First, our increasing awareness of the size of the universe and of the number of stars leads many to think that there must be other inhabitable planets. Second, if the universe is indeed "anthropic", that is, finely tuned for the production of life, then we should not consider our existence to be a mere accident, an exception, in an otherwise lifeless universe. Finally, if Linde's or any other theory that predicates either vastly many universes or an infinite number of universes is true, then despite the extremely small likelihood of anthropic sets of laws, there are likely to be (or would have to be in an infinite and varied ensemble of universes) others suited to the production of life.

These speculations, I believe, can have a healthy effect on Christian theology. In particular, they force us to "de-anthropocentrize" our theology; that is, they force us to recognize that it is not *Homo sapiens* per se that represents the goal of God's creation, but rather it is creatures of any sort who have the sensitivity and intelligence to become aware of God's love and the freedom and moral sensibility to respond appropriately.

While the ancient Greek concept of a hierarchy of beings is no longer tenable, current understandings of the relations among the sciences have provided a new conceptual structure – the hierarchy of complex systems. This hierarchy, too, runs from the inorganic to the organic, to the sentient, and finally to the conscious and intelligent. Humans here, too, occupy the highest level among earthly creatures. In light of evolutionary biology and ecology our dependence on the lower levels is clear, and so one can still say that in some sense the lower levels serve the purposes of the higher. However, there is no justification here for human exploitation of nature.

The topic of God's purpose or goal in creating is the point at which the doctrine of creation is tied to eschatology – that is, Christian teachings about the end of the world. The attempt to reconcile scientific pictures of the end of the universe with hope for human survival has led to the development of what Freeman Dyson calls "physical eschatology". Dyson's theory postulates an open universe that continues to expand and cool forever. It also depends on accepting the premise that a living creature is a type of organization that is capable of information processing. Given this definition of life, he argues that life can continue, throughout the universe, without the conditions needed for terrestrial biochemistry.[33]

Robert Russell notes that there is a dissonance between theories such as Dyson's and Frank Tipler's[34] on the one hand and traditional Christian eschatology on the other. Christian hope for the future is not reducible to unending life, but has to do rather with eternal life – "the full reality of divine time without separations and divisions, weeping and death".[35] In addition, Dyson's theory sees intelligence – in fact, mere information processing – as the attribute that makes humans valuable. As I have noted previously, traditional Christian teaching emphasizes love, not intelligence, as that which is central to God's creative purposes.

Russell points out, though, that theories such as Dyson's can remind theologians that eschatology pertains to the whole of the cosmos, not just humans. These theories should prompt us in the church, he says, "to rethink the cosmological implications of just what is at stake if we claim… that the groaning of *all* nature will be taken up in and healed by the transfiguration of the universe which has already begun with the Resurrection of Christ".[36]

Conclusion

So I claim that if you pick up any reference work on the doctrine of creation and read the list of topics discussed in traditional accounts you will find that each of these issues is being raised afresh by contemporary science. Did the universe have a true beginning in time? If so, what occurred before the beginning of the universe? Does it even make sense to speak of its beginning "in time" or is time itself an aspect of the universe? What, really, is "nothing?" Is it true that something can come out of nothing? Does the universe have a purpose and do humans have a special place in it? Is suffering necessary? Could the universe have been better than it is? What is the source of order in the universe? What is to be the fate of the universe and of the human race?

Notice that my interest here is has not been what is called natural theology. That is, it has not been my purpose in this paper to argue that theological positions can be grounded in scientific results.[37] Rather, I hope to have made three points. First, theologians cannot ignore traditional cosmological questions in treating the doctrine of creation. Second, current debates about the significance of recent scientific theories provide new conceptual resources for the theological debate. Third, our contemporary scientific culture is hungry for answers to the big questions – questions bearing on the meaning of life. The door is open to argue for the relevance of traditional Christian teaching, in an age when the relevance of religion is as much in question as its truth.

NOTES

[1] Julian Hartt, "Creation and Providence", in Peter C. Hodgson and Robert H. King, eds., *Christian Theology: An Introduction to Its Traditions and Task* (Philadelphia, Penn.: Fortress Press, 1982), pp. 115-140, quotation p. 115.

[2] Michael J. Buckley, S. J., *At the Origins of Modern Atheism* (New Haven, Conn.: Yale University Press, 1987) cf. John B. Webster, "Creation, Doctrine of", in Alister McGrath, ea., *The Blackwell Encyclopedia of Modern Christian Thought* (Oxford: Blackwell, 1993), pp. 94-97.

[3] A typical example of the rejection of theological accounts is Daniel Dennett, *Darwin's Dangerous Idea: Evolution and the Meaning of Life* (London: Penguin, 1995).

[4] See, for example, Phillip E. Johnson, *Darwin on Trial* (Downers Grove, Ill.: InterVarsity Press, 1981).

[5] Friedrich Schleiermacher, *The Christian Faith,* English trans. of the 2nd German edition (Edinburgh: T. & T. Clark, 1928), p. 76.

[6] Langdon Gilkey, "Creation", in Donald W. Musser and Joseph L. Price, eds., *A New Handbook of Christian Theology* (Nashville, Tenn.: Abingdon Press, 1992) pp. 107-113, quotation p. 108.

[7] Karl Rahner, *Foundations of Christian Faith: An Introduction to the Idea of Christianity* (New York: Seabury Press, 1978).

[8] Cf. Webster, *op. cit.;* and James Wm. McClendon, Jr., *Doctrine: Systematic Theology Volume 2* (Nashville, Tenn.: Abingdon Press, 1994), Chap. 4.

[9] *Contra Gent.,* 2, 38; *S.T.* Ia, 46, 2. Cf. Frederick Copleston, S. J., *A History of Philosophy,* Vol. 2. (New York: Doubleday, 1962), Chap. 36.

[10] 2 *Sent., I, I, I, 2,* 3; cf. Copleston, *op. cit.,* Chap. 28.

[11] Paul Davies, *The Mind of God: The Scientific Basis for a Rational World* (New York: Simon & Schuster, 1992), p. 46.

[12] Ernan McMullin, "How Should Cosmology Relate to Theology?, in A. R. Peacocke, ea., *The Sciences and Theology in the Twentieth Century* (Notre Dame: University of Notre Dame Press, 1981) pp. 1757, quotation on p. 39. However, I would not rule out use of Big-Bang cosmology as support for a theological research program that takes its *primary* support from its own proper theological data. See my *Theology in the Age of Scientific Reasoning* (Ithaca, New York: Cornell University Press, 1990) for an account of empirical support for theological research programs.

[13] See Robert J. Russell, "Cosmology from Alpha to Omega", *Zygon,* 29, 4 (1994), pp. 557-77.

[14] See Davies, *op. cit., p.* 70.

[15] See C. J. Isham, "Quantum Theories of the Creation of the Universe", in Robert J. Russell, Nancey Murphy, and C. J. Isham, eds., *Quantum Cosmology and the Laws of Nature: Scientific Perspectives on Divine Action,* Vatican Observatory and Center for Theology and the Natural Sciences, (Vatican City State and Berkeley, 1983), pp. 49-89.

[16] Davies, *op. cit., p.* 67.

[17] Stephen W. Hawking, *A Brief History of Time* (London: Bantam Books, 1988), p. 141.

[18] Anne M. Clifford, "Creation", in Francis Schussler Fiorenza and John P. Galvin, eds., *Systematic Theology: Roman Catholic Perspectives,* Vol. 1 (Minneapolis, Minn.: Fortress Press, 1991) pp. 193-248; and McClendon, *op. cit.*

[19] Russell, "Cosmology from Alpha to Omega", p. 321.

[20] Wim Drees, "Beyond the Limitations of Big Bang Theory: Cosmology and Theological Reflection", *CTNS Bulletin,* 8, 1 (1988), pp. 1-15, quotation p. 6.

[21] Robert J. Russell, "Finite Creation without a Beginning: The Doctrine of Creation in Relation to Big Bang and Quantum Cosmologies", in Russell et al., eds., *Quantum Cosmology and the Laws of Nature,* pp. 293-329.

[22] *Ibid.*, p. 325.

[23] *S.T.* Ia 44, 4.

[24] John Barrow and Frank J. Tippler, *The Anthropic Cosmological Principle* (Oxford: Clarendon Press, 1986).

[25] John Hick, *Evil and the God of Love* (Great Britain: Macmillan, 1966), p. 19.

[26] Robert John Russell, "Entropy and Evil", in *Zygon, 19,* 4 (1984), pp. 449-468, quotation p. 457.

[27] Nancey Murphy and George F. R. Ellis, *On the Moral Nature of the Universe: Theology, Cosmology, and Ethics* (Minneapolis, Minn.: Fortress Press, 1996), pp. 243-49.

[28] About $1/(10^{10})^{123}$ of the entire volume available. See Mark William Worthing, *God, Creation, and Contemporary Physics* (Minneapolis, Minn.: Fortress Press, 1996), p. 43.

[29] Davies, *op. cit.,* p. 81.

[30] See William R. Stoeger, "Contemporary Physics and the Ontological Status of the Laws of Nature", in Russell et al., eds., *Quantum Cosmology and the Laws of Nature,* pp. 209-34.

[31] Nancey Murphy, "Divine Action in the Natural Order: Buridan's Ass and Schrodinger's Cat", in Robert John Russell, Nancey Murphy, and Arthur R. Peacocke, eds., *Chaos and Complexity: Scientific Perspectives on Divine Action,* Vatican Observatory and Center for Theology and the Natural Sciences (Vatican City State and Berkeley, 1995), pp. 325-358.

[32] Hartt, *op. cit.,* pp. 119-20.

[33] Freeman J. Dyson, *Infinite in All Directions* (New York: Harper and Row, 1988), Chap. 6.

[34] Frank J. Tippler, *The Physics of Immortality: Modern Cosmology, God, and the Resurrection of the Dead* (New York: Doubleday, 1997).

[35] Russell, "Cosmology from Alpha to Omega", p. 571.

[36] *Ibid.*, p. 572.

[37] I have argued elsewhere that such results can be used as supporting data for theological research programs whose primary data come from scripture, history, and religious experience. See Murphy, "Evidence of Design in the Fine-Tuning of the Universe", in Russell et al., *Quantum Cosmology,* pp. 407-448; and Murphy and Ellis, *op. cit.*

International Symposium on Astrophysics Research and on the Dialogue between Science and Religion, Vatican Observatory, 2003
C. D. Impey & C. E. Petry, editors

Life in the Universe: An Astrobiological Perspective

Lynn J. Rothschild
NASA Ames Research Center

Introduction

For millennia humans have wondered, "Where do we come from?" and "Are we alone?" and "What is our future?" These questions have been explored using philosophical, religious and scientific perspectives. The diverse life forms on Earth apparently had a single origin. Throughout history, looking upward provoked questions regarding the nature and role of stars, the Moon, comets, meteorites and other celestial phenomena. During the day the sky was dominated by the Sun, which warmed the Earth, and living creatures depended utterly on the watery world that they inhabited. Could similar worlds exist elsewhere? Or, were humans – and indeed, all life on Earth – isolated and unique in the universe?

Astrobiology is a newly created discipline that seeks to answer these fundamental questions utilizing modern scientific and technological tools. Here we look at the nature of life in the universe from an astrobiological perspective – one that provides a scientific foundation for philosophical and religious speculations. With the advent of astrobiology, there has been an effort to understand our place in the universe in a concerted, scientific way using the latest technologies from fields as diverse as cosmochemistry and molecular biology, astronomy, and remote sensing.[1] What follows is a summary of the current state of knowledge in our quest to answer the following fundamental questions:

1. How does life originate in the universe?
2. How does life evolve?
3. How extensively is life distributed in the universe?

This paper will provide an overview of these three aspects of astrobiology. What do we know about how life arises in the universe? How does life evolve from simple to complex forms? Where is life found on Earth and on what range of sites is life found beyond Earth? It would be incredible if, in the short history of scientific inquiry, definitive answers had been obtained to any of these three questions. We may never know the complete answers. Thus, the goal of this paper is simply to frame these questions, and to summarize the current state of the research that is attempting to address them.

From the Big Bang to Life

Cosmologists tell us that the universe arose between eleven and fourteen billion years ago.[2] Recent data on the cosmic microwave background radiation using NASA's Wilkinson Microwave Anisotropy Probe (WMAP) pinpoint the age to an amazingly precise 13.7 Ga (billion years ago).[3]

In the first seconds after the Big Bang, our universe passed through a period of time when the electromagnetic and weak nuclear forces became differentiated and quarks formed. This was followed by the formation of protons and neutrons, leading ultimately to the formation of the light elements – hydrogen, deuterium, helium and lithium – all within the first three minutes.[4] After another 300,000 years, neutral atoms appeared in abundance, plentiful enough to begin to coalesce by gravity into gas clouds. The gas clouds later formed into stars, which fashioned themselves into large concentrations called galaxies. The rest of the elements formed from stars, either through nucleosynthesis – the process whereby stars produce energy by the fusion of light elements into heavier ones – or during a supernova at the end of a massive star's life. This stellar production of elements includes the so-called biogenic elements – notably carbon, oxygen, nitrogen, phosphorus, magnesium, and sulfur – which are required, along with hydrogen, for life. As Figure 1 shows, the development of intelligent life on Earth is a relatively recent event in the history of the universe and in the history of the Earth.

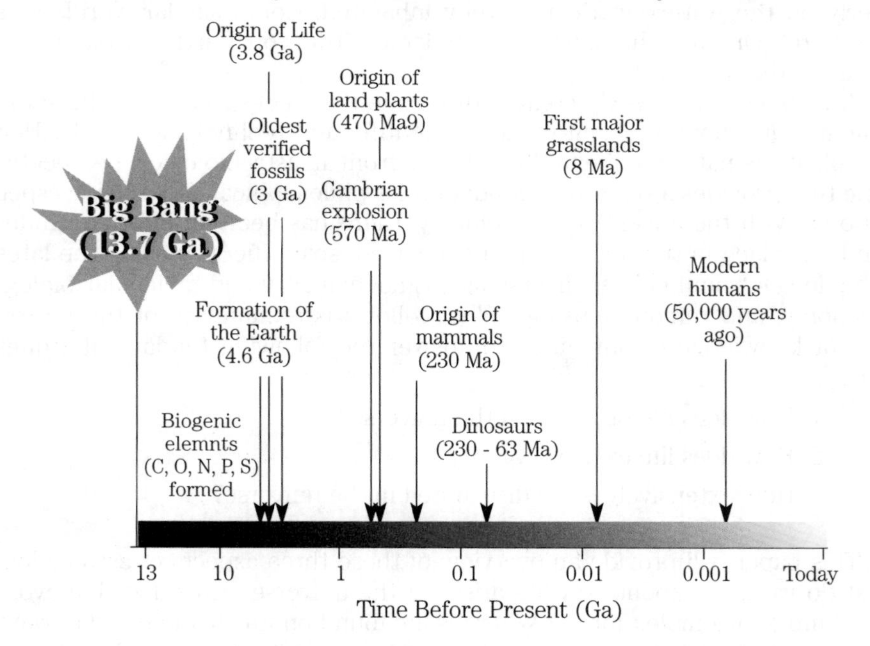

Figure 1. An overview of the evolution of the universe, from the Big Bang to humans. The scale is in billions of years before present (Ga). Note that the scale is logarithmic, which acts to compress the long interval of time on Earth before humans evolved. All dates are approximate.

The Definition of Life

At some point, about 4 billion years ago or even earlier, the most sophisticated process on Earth – chemistry based primarily on the biogenic elements – gave rise to what we recognize as life. But what do we mean by "life?" It is relatively simple to describe the universal features of life on Earth, as life apparently had a common ancestor which was the root of the genetic tree of all living organisms on Earth. With the exception of some viruses and priors, all life uses DNA as a genetic code, is partitioned from its environment by a lipid membrane, and shares the same basic chemical and metabolic attributes. Such a descriptive definition is typified by Sinsheimer,[5] who defined life as having the following characteristics:

- A flexible, self-made bounding membrane.
- Coordinated groups of specific catalysts (usually proteins).
- The molecular machinery to convert an external source of energy into forms suitable for driving reactions within the cell.
- A hereditary system of information storage (i.e. DNA).
- A method of cell division that accurately provides each new cell with a full set of hereditary instructions.
- Interlocking control systems that regulate the varied functions to allow flexibility and adaptation to changing environment.
- The provision of means for defence.

A definition of life based on thermodynamics has also been popular. Erwin Shrödinger's discourse, based on his 1943 lectures in Dublin, is the best known of these:

> When is a piece of matter said to be alive? When it goes on "doing something", moving, exchanging material with its environment, and so forth, and that for a much longer period than we would expect an inanimate piece of matter to "keep going" under similar circumstances. When a system that is not alive is isolated or placed in a uniform environment, all motion usually comes to a standstill very soon. ...After that the whole system fades away into a dead, inert lump of matter. A permanent state is reached, in which no observable events occur. The physicist calls this the state of thermodynamical equilibrium, or of "maximum entropy". Practically, a state of this kind is usually reached very rapidly.[6]

A novel approach is to examine science fiction, a genre in which the writers must have a feeling for what the audience will accept as being "alive". Discussions with Andre Bormanis, the science advisor for the popular U.S. television series *Star Trek,* revealed that four criteria are commonly used to indicate that an alien creature is indeed alive. Gene Rodenberry – the creator of the series – realized early on that if a creature has eyes, it is easy for viewers to accept it as alive. Second, if the creature is sentient, we recognize it as alive. Sentience is measured by the Turing test; that is, if the subject is unable to distinguish whether he is talking to a computer or a person, the being is recognized as living. During the series *Star Trek: The Next Generation,* an episode entitled "The Measure of a Man" probed the question of whether an apparently sentient computer called Data, a robot that functioned as a member of the crew, was actually alive. Third, we use behavior as a discriminator of living versus non-living. In *Star Trek: Deep Space 9,* a creature was introduced that behaved like a puppy although it was a software controlled mechanical device. In the original *Star Trek* series there was an episode called "Devil in the Dark" in which a rock that was really a silicon-based life form behaved like a mother trying to protect her young. Finally, entities that evolve are often considered to be alive. In the *Star Trek: The Next Generation* episode called "Evolution", particles called nannites were found that behaved and evolved collectively.

The role of reproduction and replication is also ambiguous when trying to define life. Computer programs can be written with instructions for selfreplication, but most people would probably not consider programs to be alive. What about computers in the form of robots that produce other robots? An element of variation could be introduced into the replication process to mimic the heritable variation we see in the natural world. If we are required to assist with the replication of a computer or robot, does that mean it is not alive? This issue becomes clouded when the concept of parasites, especially viruses, is introduced. Viruses do not have the machinery with which to perform their own replication, and thus must commandeer the host cell's metabolic machinery. These fictional examples are designed to erode our confidence that we can distinguish between sentient beings, robots, and simple organisms that display complex behavior.

All of this speculation appears to bring us no closer to a concrete definition of life. We realize that this issue is extremely difficult to resolve definitively, particularly in the computer age. It is unlikely that any definition can be found that excludes computer-based life. Reluctantly, we conclude that the privilege to define life is a spoil that goes to the victor, which on Earth today clearly means humans. We are forced to be arbitrary in what we admit to the "club of life". However, we must also realize that if there are other sentient races in the universe, they may have a different definition, perhaps one that doesn't include humans.

The Necessary Conditions for Life

Of all the simple questions about the natural world, "How does life originate?" is the one scientists are least able to answer satisfactorily. On one hand, we are able to take apart a cell and analyze its chemical mechanisms in exquisite detail: we can describe and manipulate it, and we can synthesize its components. On the other hand, the reverse experiment has not yet been conducted in the lab: we cannot generate a living organism from simple chemical ingredients. With the possible exception of viruses[7] – if we even acknowledge viruses as living – we have yet to create life *de novo*. Because the wealth of information about cellular components is so great, one has the feeling that the answer is near at hand. Perhaps life only emerges if the right components are mixed under the right conditions, which suggests that the problem is one of finding the right recipe with the ingredients we have identified. But even though we are tantalizingly close to a complete understanding of simple living organisms, we have made little progress in creating a self-sustaining being. This suggests that we are still profoundly ignorant of some of the essential mechanisms of life.

We know that there are certain minimal conditions needed for life: the presence of liquid water, the biogenic elements, a source of energy and aufficient time. The assumption is that all life in the universe is going to be based on organic carbon as life on Earth is. Why should this be so?

Carbon is the fourth most common element in the universe. It is capable of forming a vast array of compounds ranging from methane to DNA. It is uniquely able to form large and stable molecules, thus it is the perfect storage medium for the genetic code. In addition, atomic carbon and simple organic compounds with up to 13 atoms have been detected in interstellar space, including amino acids and nucleotide bases.

Even though these minimal conditions seem quite restrictive, and the assumptions behind them cannot be completely just)fied, the last two decades have shown that all of the criteria for the development of carbonbased life are met in many places throughout the universe. This represents perhaps the most stunning advance in astrobiology in the last two decades.

The European Space Agency's Infrared Space Observatory (ISO) mission, which operated from November 1995 to May 1998, revealed far more water than we had imagined. The water vapor made each day in the gas cloud in the Orion nebula is thought to be enough to fill the Earth's oceans 60 times over.[8] Water is trapped in planets as they form, and is delivered to planets by comets, though the relative importance of each is still debated. Similarly, three decades ago we thought interstellar space was truly empty. Today we know that it is populated with traces of organic compounds, the building blocks of life.[9, 10, 11, 12, 13]

Stars are an obvious source of energy, but we now know that tidal flexing between a moon and its parent planet can occur, generating enough energy to melt water even at distances far beyond the Sun's warming rays. For example, tidal heating resulting from the interaction between Jupiter and its moon Europa is aufficient to allow liquid water under the ice-covered surface of Europa.

Finally, we need to define "sufficient time". At a minimum we need enough time for the biogenic elements (carbon, oxygen, nitrogen, phosphorus, magnesium) to build up through stellar processes to a significant fraction of solar abundance. Thus, there must be a sufficient time for the completion of at least one generation of stars after the Big Bang, and for the beginning of the subsequent generation. Biologically, time is needed for the production of organic molecules and their organization into selfperpetuating bodies. How long does this take? The Earth was subject to a heavy bombardment by extraterrestrial bodies left over from the formation of the solar system, such as meteorites and comets, until about 3.8 billion years ago. Some of these impacts may have been energetic enough to have sterilized the planet; this is known as the "impact frustration" of life.[14] Yet multiple lines of evidence suggest that life appeared nearly coincident with the end of the late bombardment period. Life on Earth could have arisen more than once.

What can we conclude from the available data? Life arose early on Earth and was able to survive the impacts by hiding in various refuges, such as in the deep subsurface. Life probably arose earlier on a less hostile body such as Mars, and may have then traveled to the Earth. At this time, Mars was likely to have been warmer and wetter. Because of its small size, it was unable to retain an atmosphere and thus lost both its potential greenhouse effect and much of its water inventory. Lastly, we infer that under the proper conditions, life can arise very rapidly indeed. The jury is still very much out on whether we know either the necessary or the aufficient conditions for the origination of life.

The Evolution of Life on Earth

Where did life originate on Earth? Did it arise in the soil or the upper atmosphere, or tropical lagoons, or in oceanic hydrothermal vents? The list of possible enviroments is extensive. We are still not certain, but molecular sequence data hint that the last common ancestor for life may well have been a thermophile, that is, an organism that preferred to live at high temperatures. Possibly this was because it was the only creature able to survive a nearly sterilizing impact.

In fact, we are not even sure if life began in the first instance on Earth. Given what we know about the evolution of the Martian climate, more time was available for life to begin on Mars. We also know from meteorites – such as the now famous Allen Hills 84001 – that material can be ejected from the surface of Mars and make its way to the Earth. The potential for this sort of transfer, known as pansperrnia, is now well within the scope of scientific experimentation. For example, NASA and the European Space Agency (ESA) have had a series of missions where organisms were sent into space and subsequently exposed to space vacuum and radiation. Most recently, ESA has been responsible for the BioPan experiments where it was found that spores, as well as halophilic (salt-loving) microbes, were able to survive in the vegetative state.[15]

During the last part of the 20th century, there was a lively debate regarding the course of evolution once life originated. The principals have been the late Stephen Jay Gould of Harvard University, and Simon Conway Morris of the University of Cambridge. Gould maintained in a series of essays and books[16] that if "the tape" of the history of the Earth were replayed, the course of evolution and thus the diversity of life would be unpredictable and unrecognizable. In contrast, Conway Morris[17] argued that evolutionary principles exist, continually shaping the way life evolves as the tape plays. Gould may be right in the details, but Conway Morris is likely to be right in the broad outline. I will enter the fray by outlining several principles of evolution and seeing where they lead. The coverage is not meant to be exhaustive; the transition of life from water to land and back, the evolution of human skin color, and the evolution of locomotion would all have made equally good examples of convergent evolution.

Key Events in Evolution: Cellular Organization

Just as atoms are the basic units in chemistry, cells are the basic units for life on Earth. There are two distinct organizational plans: one in which the genetic material is enclosed in a membrane that creates the cell's nucleus, and another in which the genetic information is in the cytoplasm but not bound within a membrane. A membrane-bound nucleus is found in all members of the domain Eukarya, a group that includes organisms as seemingly diverse as protists, fungi, plants, and animals. The remaining domains are equally or perhaps more genetically diverse: Archaea and Bacteria. Both domains appear similar under the microscope, but they have molecular differences so profound that Woese and Fox[18] placed them in separate domains – a separation that is widely accepted today.

Multicellular organization is a deceptively simple concept, easy to regard as somehow an "advanced" manifestation of life. Multicellularity can take the form of filaments or other aggregates of cells that may or may not show cooperation or differentiation in their function. Alternatively, multicellularity can be so rigid as to dictate the role and fate of each individual cell. Each cell must respond in ways that will help the organism as a whole to produce offspring so that its genes will be passed on to the next generation.

However, life on Earth displays astounding variety of both function and form, and two other situations might well qualify as multicellular. First, many microbes as a whole come together so that they are touching or in contact through chemical signals, and through these means they interact and function as a group. The fruiting bodies created by the myxobacteria are well known, as are those created by the cellular slime mold *Dictyostelium.* Even microbial mats, which are formed from an intimate association of microbes, have been considered to be multicellular "organisms".[19]

Second, there are many so-called "unicellular" organisms that have more than one cell type in their life cycle. Many protists have two or more stages in their life cycles, each with discrete morphologies, capabilities, and associated genetics. The frog gut commensal *Opalina* – with nearly 30 stages in its life cycle – is a dramatic, but by no means unusual example. I would suggest that such organisms also be called "multicellular" – but in a temporal rather than in a spatial sense. If this is the case, multicellularity may be the rule for life on Earth rather than the very visible exception.

There are members of all three domains that have become multicellular, from kelp to whales to sponges to the fruiting bodies of the bacterium *Myxococcus* (see Figure 2). Multicellularity is a convergent evolutionary trait and thus arose repeatedly. The best way to think of multicellularity is as one of many adaptive strategies. Multicellularity has the advantage of increasing an organism's size, allowing for the formation of specialized structures, enhancing the ability to prey on other organisms (and avoid predation), and allowing specialization of function at the cellular level. However, there are costs associated with multicellularity, including the fact that previously adaptive features such as locomotion become "altruistic" since they benefit other cells as well.[20]

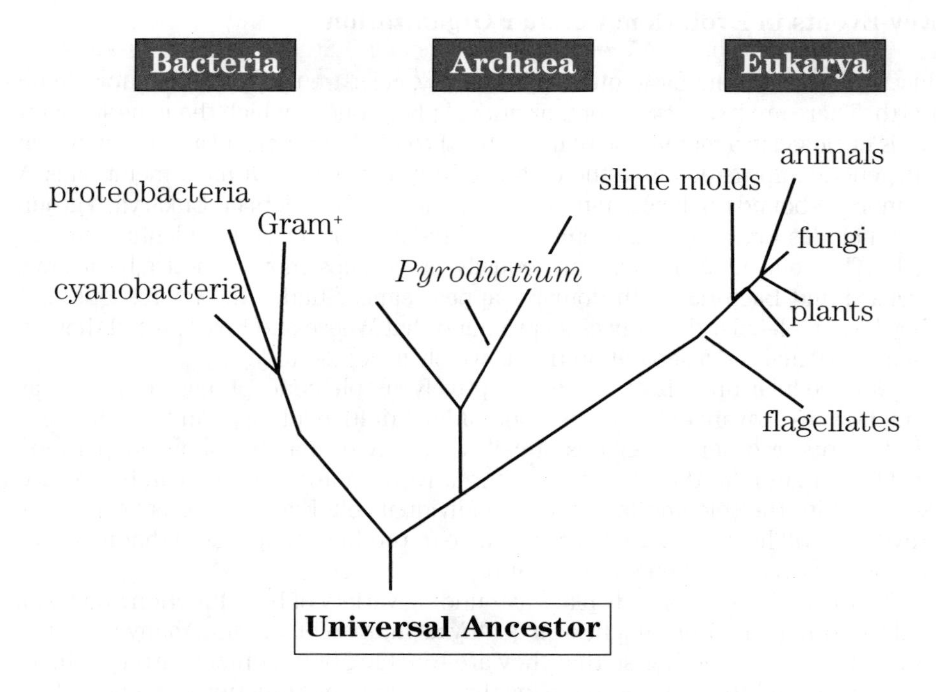

Figure 2. The distribution of multicellularity among major groups of lifeforms of Earth (taxa). Of the three domains of life, at least two have multicellular members. In most cases, there is at least some differentiation of cell types and cooperative behavior. This taxonomic distribution suggests that multicellularity arose multiple times and thus is advantageous for the long term survival of organisms.

Key Events in Evolution: Metabolism

Earlier, I argued for the assumption that the universality of life is based on organic carbon – carbon combined with hydrogen and nearly always oxygen as well. The evidence for the hypothesis is the unique ability of carbon to form long chain molecules, and the universal availability of carbon, nitrogen, hydrogen, and oxygen. Even more persuasive is the fact that organic chemistry pervades the interstellar medium – a low density, gaseous environment quite distinct from planetary surfaces.

If we assume that all life – on Earth and elsewhere in the universe – will be based on organic carbon, a basic script for metabolic evolution emerges. Organic carbon is widespread, but inorganic forms of carbon – such as graphite, diamonds, carbon dioxide, carbonate and carbon monoxide – are far more common, even on Earth. Thus, as an organic carbon-based life form begins to thrive, organic carbon itself is likely to be the nutrient that limits the growth af the organism. Selection pressure will arise for organisms that can transform inorganic to organic carbon, a process called carbon fixation and whose practitioners are called "autotrophs", or in the ecological literature, "primary producers". Indeed, this is precisely what happened on Earth. The most successful autotrophs have been those that evolved to exploit sunlight as a source of energy for this process, and can extract hydrogen from water to use as a reductant in the process called oxygenic photosynthesis. Its practitioners are found wherever there is even a modest amount of light; they are called plants, algae and cyanobacteria.

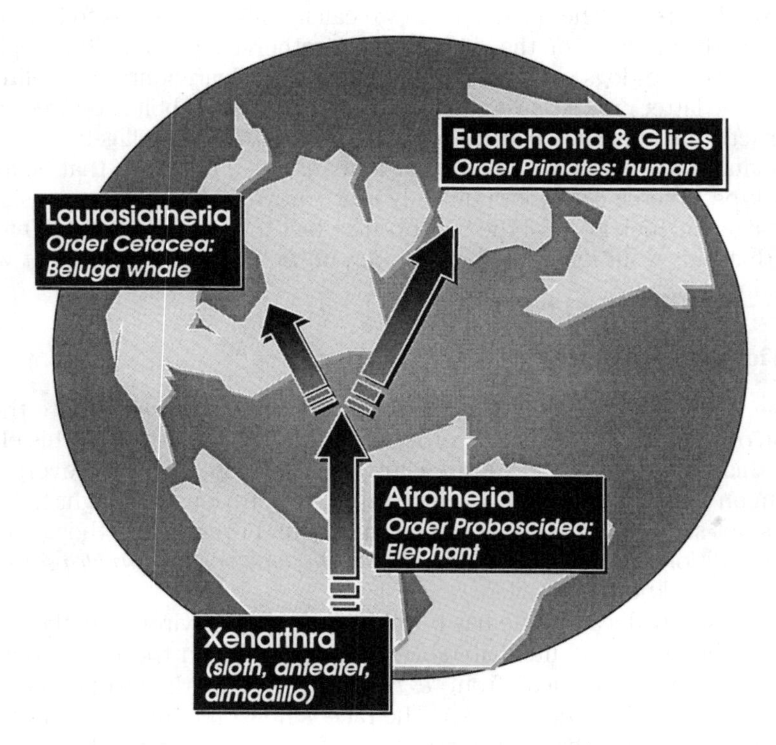

Figure 3. The schematic and geographical distribution of intelligence among placental mammals. Note that intelligent species have arisen independently in both northern taxa. The sketch shows the major continents as they would have appeared at the time of the Afrotheria and Laurasiatheria split, estimated at 111-118 million years ago. Data from Madsen et al.

Life evolves to maximize reproductive success, and economy of metabolism is an excellent strategy. Thus, the strategy of evolving to consume autotrophs will be advantageous in a variety of circumstances. In this way herbivory was born. However, the food quality of the herbivores themselves can be better than that of plants and other autotrophs, and so we see the evolution in multiple taxa of carnivores.

Key Events in Evolution: Intelligence

If prey is active, carnivory is often associated with an increase in intelligence. That is because complex behavior in reaction to changing situations requires rapid and flexible response. If we focus on the mammals, the new molecular evidence arising from the laboratories of Mark Springer and his colleagues suggests that there have been four distinct lines (or radiations) of placental mammals.[21] The resulting groups are the Xenarthra, Afrotheria, Laurasiatheria, and Euarchonta plus Glires (see Figure 3). The phenomenon we call intelligence seems to have arisen independently in three of the lines, with Afrotheria including the elephants, Laurasiatheria the dogs, whales and porpoises, and Euarchonta plus Glires the humans. The latter group also contains mice, porcupines, rabbits and tree shrews – not exactly distinguished cousins for humans in terms of intelligence.

Human dentition clearly shows that humans are omnivores, that is, able to eat any type of food. Thus, it is unlikely that carnivory alone is responsible for our intelligence. Rick Potts of the Smithsonian Institution in Washington has suggested that our evolutionary response to an unpredictable environment was to develop intelligence.[22]

How Widespread is Life?

To gauge where life might exist in the universe, it is important to note the distribution of life on Earth, in other words, what are the limits to life on this planet? Why is this important for the search for life elsewhere? The discovery of an organism on Earth which inhabits a niche that was previously thought to be barren does not suggest that such an organism is found everywhere those environmental conditions prevail. Rather, *it shows the minimum envelope for life*, that is, what we know is possible.

In the last few decades, life has been discovered in environmental extremes previously thought to be uninhabitable. The organisms in these environments were dubbed "extremophiles", from *"extremus"*, the superlative of *exter* (being on the outside), and *"philos"* (love). The race is now on to discover the "champions" – the organisms that can survive the highest temperature, the lowest pH, and other physical extremes.

The environmental extremes of interest range from physical extremes (e.g., temperature, pressure), to chemical extremes (e.g., pH, salinity) and possibly even biological extremes (e.g., population density, parasite loading). There are good physical and biological reasons to postulate that any of these variables taken to the extreme would provide obstacles for the long term development of carbon-based life forms.

For example, at high temperatures, nucleic acids and proteins denature (unfold), and thus lose function. Chlorophyll is destroyed at a temperature greater than about 73°C. Lipid membranes such as those that enclose the cells become fluid, so the integrity of the cell as a structural unit is lost. The solubility of gases decreases as temperature increases, which means that essential gases, such as oxygen for animals and carbon dioxide for plants, become scarce.

Table 1. Environmental Extremes for Life

Environmental Parameter	**High**	**Low**
Temperature	*Pyrolobus fumarii,* 113°C	Penguins, polar bears, etc. Some insects and microbes show activity below -10°C. Survival at low temperatures widespread
Radiation	*Deinococcus radiodurans*	Theoretically zero
Pressure	Some microbes, 130 MPa	
Gravity	Cells/small organisms – 10^6 G Young plants – 10 minutes at 30-40 G Rats – 15 G for 10 min (20 G is lethal) Humans – 4-5G for 10 minutes	Survival good in microgravity (e.g., the International Space Station). However, multi-generational data are limited. Bacteria have produced multiple generations. Frogs only through early tadpoles (~4 days) with no second (F2) generation. *Brassica* (plant) through F2 with some problems
Vacuum	Tardigrades, insects, microbes, seeds	
Dessication	Brine shrimp, nematodes, microbes, spores, fungi, lichens	
Salinity	Microbes such as *Halobacteriacea, Dunaliella saline* (2-5 M NaCI)	Many organisms can thrive in fresh water through adaptations to this hypoosmotic environment
Acidity (pH)	*Natronobacterium, Bacillus firmus* OF4, *Spirulina spp.* (all. pH 10.5)	*Cyanidium caldarium, Ferroplasma sp.* (both pH 0)
Oxygen Tension	Many organisms as atmospheric oxygen levels may have been nearly twice present levels 300 Ma.[23]	*Methanococcus jannaschii* (cannot tolerate O_2)
Chemical Extremes	*Cyanidium caldarium* (pure CO') *Ferroplasma acidarmanus* (Cu, As, Cd, Zn); *Ralstonia sp.* CH34 (Zn, Co, Cd, Hg, Pb)	

Yet, the Archaean *Pyrolobus fumarii* lives well above the boiling point of water, at 113°C.[24] Conversely, at low temperatures membranes become rigid, biochemical reactions slow, and ice crystals may form which can pierce cell membranes.[25] Yet, life is able to survive well below the freezing point of water, from penguins to trees to microbes and insects. Cell lines can readily be preserved even in liquid nitrogen.

The environmental ranges for life on Earth are summarized in Table 1 (data from Rothschild and Mancinelli). This gives the n-dimensional space within which we *know* life can survive. This envelope overlaps with the environmental conditions of extraterrestrial locations such as Mars and Europa. Much of the data have been obtained in the last 25 years. Thus, we see from the study of extremophiles that the search for life beyond Earth is well motivated and will no doubt continue.

Conclusions

It is an enormous challenge to elucidate the origin, evolution and distribution of life in the universe, but the importance of these questions to mankind's collective psyche has encouraged an interdisciplinary approach spanning thousands of years. Beginning in the last decade of the 20th century, the modern tools of science have been brought to bear on these questions in the new discipline of astrobiology.

Recent gains in knowledge of the history of life on Earth encourage optimism that life might be found elsewhere in the solar system, and beyond. Carbon and water are widely distributed in the cosmos, and simple living organisms have adapted to a wide range of physical conditions on Earth – a range that encompasses the likely conditions on many extrasolar planets and moons. Regardless of whether intelligence is a convergent or contingent outcome of biological evolution, the universe is likely to contain a sufficient diversity of biological experiments that intelligence has been an outcome for more than one of them. In all of this speculation, the possibility of nonbiological forms of life cannot be ruled out.

Even though large gaps in our knowledge remain, we are drawn to consider the implications of progress in the field of astrobiology. If any kind of extraterrestrial life is found, it will mark a new phase in the Copernican Revolution – demonstration that not even our biological function is special. If intelligent life is found beyond Earth, the implications for organized religion will be enormous. The challenge for the future is two-fold: to continue in this quest, while using the knowledge gained from this research wisely.[26]

NOTES

[1] L. J. Rothschild, "Astrobiology", *McGraw Hill Encyclopedia of Science & Technology,* 2002, pp. 21-24.

[2] R. Cayrel, V. Hill, T. C. Beers, B. Barbuy, M. Spite, F. Spite, B. Plez, J. Andersen, P. Bonifacio, P. François, P. Molaro, B. Nordstrom and F. Primas, "Measurement of Stellar Age from Uranium Decay", *Nature,* 2001, Vol. 409, pp. 691-692; L. M. Krauss & B. Chaboyer, "Age estimates of globular clusters in the Milky Way: Constraints on Cosmology", *Science,* 2003, Vol. 299, pp. 65-69.

[3] See http://www.gsfc.nasa.gov/topstory/2003/0206mapresults.html for the recent WMAP results.

[4] P. J. E. Peebles, D. N. Schramm, E. L. Turner and R. G. Kron, "The Evolution of the Universe", *The Scientific American Book of Astronomy* (New York: The Lyons Press, 1999), pp. 241-250.

[5] R. L. Sinsheimer, "What is Life? A Closer Look", *Engineering & Science,* 1996, No. 3, pp. 35-39.

[6] E. Schrodinger, *What is Life: The Physical Aspect of the Living Cell* (London: Cambridge University Press, 1951). Based on lectures delivered under the auspices of the Institute at Trinity College, Dublin, in February 1943.

[7] J. Cello, A.V. Paul and E. Wimmer, "Chemical Synthesis of Poliovirus cDNA: Generation of Infectious Virus in the Absence of Natural Template", *Science,* 2002, Vol. 297, pp. 1016-1018.

[8] M. Harwit, D.A. Neufeld, G. J. Melnick and M. J. Kaufman, "Thermal Water Vapor Emission from Shocked Regions in Orion", *The Astrophysical Journal Letters,* 1998, Vol. 497, pp. 105-108.

[9] S. A. Sandford, "The Inventory of Interstellar Materials Available for the Formation of the Solar System", *Meteoritics and Planetary Science,* 1996, Vol. 31, pp. 449-476.

[10] L. J. Allamandola, S. A. Sandford, and M. P. Bernstein, "Interstellar/ Precometary Organic Material and the Photochemical Evolution of Complex Organics", in C. B. Cosmovici, S. Bowyer and D. Werthimer, eds., *Astronomical and Biochemical Origins and the Search for Life in the Universe* (Bologna: Editrice Compositori, 1997), pp. 23-49.

[11] S. A. Sandford, L. J. Allamandola, and M. P. Bernstein, "Organic Chemistry: From the Interstellar Medium to the Solar System", *Origins, ASP Conference Series,* Vol. 148 (San Fransisco: Astronomical Society of the Pacific, 1998).

[12] M. Bernstein, L. J. Allamandola and S. A. Sandford, "Life's Far-Flung Raw Materials", *Scientific American, 1999,* pp. 42-49.

[13] G. C. Sloan, T. L. Hayward, L. J. Allamandola, J. D. Bregman, B. DeVito and D. M. Hudgins, "Direct Spectroscopic Evidence for Ionized Polycyclic Aromatic Hydrocarbons in the Interstellar Medium", *The Astrophysical Journal Letters, 1999,* Vol. 513, pp. 65-68.

[14] K. J. Zahnle and N. H. Sleep, "Impacts and the Early Evolution of Life", in P. J. Thomas, C. F. Chyba and C. P. McKay, eds., *Comets and the Origin and Evolution of Life* (New York: Springer, 1997), pp. 175208.

[15] G. Horneck, "Could Life Travel Across Interplanetary Space? Panspermia Revisited", in L. Rothschild and A. Lister, eds., *Evolution on Planet Earth: The Impact of the Physical Environment* (London: Academic Press, 2003).

[16] S. J. Gould, *Wonderful Life: The Burgess Shale and the Nature of History* (New York: W. W. Norton, 1989), p. 323.

[17] S. Conway Morris, *The Crucible of Creation: The Burgess Shale and the Rise of Animals* (Oxford: Oxford University Press, 1998), p. 272.

[18] C. R. Woese and G. E. Fox, "Phylogenetic Structure of the Prokaryotic Domain: The Primary Kingdoms", *Proceedings of the National Academy of Science, USA,* 1997, Vol. 74, pp. 5088-5090.

[19] J. A. Shapiro and M. Dworkin, *Bacteria as Multicellular Organisms* (New York: Oxford University Press, 1997).

[20] R. Michod and D. Roze, "Cooperation and Conflict in the Evolution of Multicellularity", *Heredity,* 2001, Vol 86, pp. 1-7.

[21] O. Madsen, M. Scally, C. J. Douady, D. J. Kao, R. W. DeBry, R. Adkins, H. M. Amrine, M. J. Stanhope, W. W. de Jong and M. S. Springer, "Parallel Adaptive Radiations in Two Major Clades of Placental Mammals", *Nature,* 2001, Vol. 409, pp. 610-614.

[22] R. Potts, "Variability Selection in Hominid Evolution", *Evolutionary Anthropology,* 1998, Vol. 7, pp. 81-89; "Environmental Variability and Its Impact on Adaptive Evolution, with Special Reference to Human Origins", in L. Rothschild and A. Lister, eds., *Evolution on Planet Earth: The Impact of the Physical Environment* (London: Academic Press, 2003).

[23] R. A. Berner, "Atmospheric Oxygen over Phanerozoic Time", *Proceeding of the National Academy of Science, USA, 1999,* Vol. 96, pp. 10955-10957.

[24] E. Blochl, R. Rachel, S. Burggraf, D. Hafenbradl, H. W. Jannasch and K. O. Stetter, "*Pyrolobus fumarii,* gen. and sp. nov., Represents a Novel Group of Archaea, Extending the Upper Temperature Limit for Life to 113 °C", *Extremophiles,* 1997, Vol. 1, pp. 14-21.

[25] A. Clarke, "Evolution and Low Temperatures", in L. Rothschild and A. Lister, eds., *Evolution on Planet Earth: The Impact of the Physical Environment* (London: Academic Press, 2003).

[26] I would like to thank Emily Holton for data on gravity tolerance, and Max Bernstein and Lou Allamandola for references on interstellar chemistry. Most of all, I wish to thank George Coyne and Chris Impey for encouragement, patience, and the invitation to participate in this conference and book.

International Symposium on Astrophysics Research and on the Dialogue between Science and Religion, Vatican Observatory, 2003
C. D. Impey & C. E. Petry, editors

Science and Buddhism: At the Crossroads

Trinh Xuan Thuan
University of Virginia

The Outer World of Science and the Inner World of Buddhism

As an astrophysicist studying the formation and evolution of galaxies, my work often raises questions about matter, space and time. As a Vietnameseborn raised in the Buddhist tradition, whenever I came up against these concepts, I couldn't help wondering about how Buddhism would have dealt with them and how its view of reality compares to the scientific viewpoint. But I wasn't sure whether such questions even made sense. I was familiar with and appreciated Buddhism as a practical philosophy that provides a guide for self-knowledge, spiritual progress and becoming a better human being. Thus, as far I knew, Buddhism was primarily a pathway leading to Enlightenment, a contemplative approach with an essentially inward gaze, in contrast to science's outward look.

What's more, science and Buddhism have radically different methods of investigation of reality. In science, intellect and reason play the leading roles. By dividing, categorizing, analyzing, comparing and measuring, scientists express natural laws in the highly abstract language of mathematics. Intuition is not absent in science, but it is only useful if backed up by a coherent mathematical formulation and validated by observation and analysis. On the other hand, it is intuition – or rather inner experience – that plays the leading role in the contemplative approach. Rather than breaking up reality, it aims to understand it in its entirety. Buddhism has no use for measuring apparatus and does not rely on the sort of sophisticated observations that form the basis of experimental science. Its statements are more qualitative than quantitative. So I was far from sure that there would be any point confronting science and Buddhism. I was afraid that Buddhism would have very little to say about the nature of phenomena, because this is not its main interest, whereas such preoccupations lie at the heart of science.

In the summer of 1997, I met the French Buddhist monk Matthieu Ricard at the University of Andorra where we were both giving lectures. He was the ideal person to discuss these issues with: he was trained as a scientist (he has a doctorate in biology from the Pasteur Institute in Paris), so is familiar with the scientific method, but he is also well- versed in Buddhist texts and philosophy as he has left the scientific world to become a Buddhist monk in Nepal about 30 years ago. We had many fascinating discussions during our long walks together in the inspiring mountain scenery of the Pyrenees. Our conversations were always mutually enriching. They led to new questions, original viewpoints and unexpected syntheses that required further study and clarification, and still do so. I shall discuss below the main issues that sometimes divided us, sometimes united us. A book *The Quantum and the Lotus*[1] was born from those friendly exchanges between an astrophysicist who wanted to confront his scientific knowledge with his Buddhist philosophical origins with a Western scientist who became Buddhist and whose personal experience has led him to compare these two approaches.

At the close of our conversations, I must say that my admiration for how Buddhism analyzes the world of phenomena has grown considerably. It has thought deeply and in a profoundly original way about the nature of the world. But the ultimate goals for the pursuit of knowledge in science and Buddhism are not the same. The purpose of science is to find out about the world of phenomena. In Buddhism, knowledge is acquired essentially for therapeutic purposes. The objective is not to find out about the physical world for its own sake, but to free ourselves from the suffering that is caused by our undue attachment to the apparent reality of the external world. By understanding the true nature of the physical world, we can clear away the mists of ignorance and open the way to Enlightenment.

It is not my purpose in this paper to make science sound mystical nor to justify Buddhism's underpinnings with the discoveries of science. Science is perfectly self-sufficient and accomplishes well its stated aim without the need of a philosophical support from Buddhism or from any other religion. Buddhism is a science of the Enlightenment, and whether it is the Earth that goes around the Sun or the contrary cannot have any consequence on its philosophical basis. But because both are quests for the truth, and both use criteria of authenticity, rigor and logic to attain it, their respective views of the world should not result in an insuperable opposition, but rather to a harmonious complementarily. Werner Heisenberg has expressed this eloquently: "I consider the ambition of overcoming opposites, including also a synthesis embracing both rational understanding and the mystical experience of unity, to be the mythos, spoken or unspoken, of our present day and age".[2]

I shall discuss below the Buddhist concepts on interdependence, emptiness, and impermanence, and how they find an echo in modern science. I shall outline how Buddhism rejects the idea of an "anthropic" principle. Finally, I will conclude that science and spirituality are two complementary modes of knowledge and that they must go hand in hand so that we do not forget our humanity.

Interdependence

Buddhism and the Interdependence of Phenomena

One of Buddhism's central tenets is the "interdependence of phenomena". Nothing exists inherently, or is its own cause. An object can be defined only in terms of other objects and exists only in relationship to others. In other words, this arises because that exists. Interdependence is essential to the manifestation of phenomena. In Buddhism, the perception of distinct phenomena resulting from isolated causes and conditions is called "relative truth" or "delusion". Our daily experience makes us think that things possess a real, objective independence, as though they existed all on their own and had intrinsic identities. But Buddhism maintains that this way of seeing phenomena is just a mental construct. Rather it adopts the notion of mutual causality: an event can happen only because it is dependent on other factors. Any given thing in the world can appear only because it is connected, conditioned, and in turn conditioning; an entity that exists independently of all others as an immutable and autonomous entity could not act on anything, or be acted on itself.

Buddhism thus sees the world as a vast flow of events that are linked together and participate in one another. The way we perceive this flow crystallizes certain aspects of the non-separable universe, thus creating the illusion that there are autonomous entities completely separate from us. Thus phenomena are simply events that happen in some circumstances. This view does not mean that Buddhism denies conventional truth – the sort that ordinary people perceive or that the scientist detects with his apparatus – or that it contests the laws of cause and effect, or the laws of physics and mathematics. It simply holds that, if we dig deep enough, there is a difference between the way we see the world and the way it really is.

The most subtle aspect of interdependence concerns the relationship between a phenomenon's "designation bases" and its "designation". An object's "designation bases" refer to its position, dimension, form, color or any other of its apparent characteristics. Together, they comprise the object's "designation", a mental construct that attributes an autonomous distinct reality to that object. In our every day experience, when we see an object, we aren't struck by its nominal existence, but by its designation. Because we experience it, Buddhism does not say that the object doesn't exist. But neither does it say that the object possesses an intrinsic reality. So it concludes that the object exists (thus avoiding the nihilistic view that Westerners too often attribute mistakenly to Buddhism), but that this existence is purely interdependent. A phenomenon with no autonomous existence, but which is nevertheless not totally inexistent, can thus act and function according to the laws of causality.

Non-separability in Quantum Mechanics

A notion strikingly similar to that of Buddhism's interdependence is the concept of non-separability in quantum mechanics based on the famous thought experiment proposed by Einstein, Podolsky and Rosen (EPR) in 1935.[3] In simplified terms, the experiment goes as follows. Imagine a particle that disintegrates spontaneously into two photons A and B. The law of symmetry dictates that they will travel in opposite directions. If A goes northwards, then we will detect B to the south. It all seems perfectly normal. But that's forgetting the strangeness of quantum mechanics. Particles can also appear as waves. Before being captured by the detector, A appeared as a wave, not a particle. This wave was not localized, so that there was a certain probability that A might be found in any direction. It's only when it has been captured that A changes into a particle and "learns" that it's heading northwards. But, if A didn't "know" before being captured which direction it had taken, how could B have "guessed" what A was doing and adjusted its behavior accordingly so that it could be captured at the same time in the opposite direction? This is impossible, unless A can inform B instantaneously of the direction it has taken. As Einstein said, "God does not send telepathic signals", and there can be "no spooky action at a distance". He therefore concluded that quantum mechanics did not provide a complete description of reality, that A must "know" which direction it was going to take and "tell" B before they split up. According to him, there must be "hidden variables" and quantum mechanics must be incomplete.

Yet Einstein was actually wrong. In 1964, John Bell devised a mathematical theorem called "Bell's inequality" which could be verified experimentally if particles really did have hidden variables. In 1982, Alain Aspect and his team in Paris carried out a series of experiments on pairs of photons and found that Bell's inequality was always violated. Quantum mechanics was right and Einstein was wrong. In Aspect's experiment, photons A and B were 12 meters apart, yet B always "knew" instantaneously what A was doing, and reacted accordingly. In the latest experiment carried out by Nicolas Gisin and his team in Geneva, the photons are separated by 10 kilometers, and yet their behaviors are perfectly correlated. This is bizarre only if, like Einstein, we think that reality is cut up and localized in each photon. The problem goes away if we admit that A and B, once they have interacted with each other (the physicists describe them as "entangled") become part of a non-separable reality, no matter how far apart they are, even if they are at opposite ends of the universe. A doesn't need to send a signal to B because they share the same reality.

Quantum mechanics thus eliminates all idea of locality and provides a holistic view of space. The notions of "here" and "there" become meaningless, because "here" is identical to "there". That is what physicists call "non-separability". So phenomena do seem to be "interdependent" at the subatomic level, to use the Buddhist term.

Foucault's Pendulum and Interdependence in the Macrocosm

Another fascinating and famous physics experiment demonstrates that interdependence isn't limited to the world of particles, but applies also to the entire universe. This is the pendulum experiment carried out by Leon Foucault in 1851 to demonstrate the rotation of the Earth. We are all very familiar with the behavior of the pendulum. As time passes, the direction in which it swings changes. If the pendulum were placed at either the North or South pole, it would turn completely round in twenty- four hours. In Paris, because of a latitude effect, it takes more than twenty-four hours to make a complete turn. Foucault realized that, in fact, the pendulum always swung in the same direction, and it was the Earth that turned.

But there remains a puzzle not clearly understood to this day. The pendulum is attached to a building that is attached to Earth. The Earth carries us at some

30 km s^{-1} around the Sun, which is itself flying through space at 230 km s^{-1} in its orbit around the center of the Milky Way. Our Galaxy, in turn, is falling toward the Andromeda galaxy at 90 km s^{-1}. The Local Group of galaxies, whose most massive members are the Galaxy and Andromeda, is moving at 600 km s^{-1} under the gravitational attraction of the Virgo cluster, and of the Hydra-Centaurus supercluster. The latter is itself falling toward the Great Attractor, the mass of which is equivalent to that of tens of thousands of galaxies. All of these masses and motions are local.

Yet, the Foucault pendulum seems to disregard all of them and align itself with the rest of the universe, i.e. with the most distant clusters of galaxies known. Thus, what happens here on Earth is decided by the vast cosmos. What occurs on our tiny planet depends on all the structures in the universe. Why does the pendulum behave in that way? We don't know. Ernst Mach thought that it could be explained by a sort of omnipresence of matter and of its influence. According to him, the correlation between the plane of oscillation of the Foucault pendulum and the distant clusters of galaxies comes from the distant universe being responsible for the inertia of the pendulum, and hence of its motion, through a mysterious interaction which he did not precisely understand. Again, we are drawn to a conclusion that resembles very much Buddhism's concept of interdependence: that each part depends on all the other parts.

Emptiness: The Absence of an Intrinsic Reality

The notion of interdependence leads us directly to the idea of emptiness or "vacuity" in Buddhism, which does not mean nothingness (as often thought erroneously by Westerners), but the absence of inherent existence. Since everything is interdependent, nothing can be self-defining and exist inherently. The idea of intrinsic properties that exist in themselves and by themselves must be thrown out.

Once again, quantum physics has something strikingly similar to say. According to Bohr and Heisenberg, we can no longer talk about atoms and electrons as being real entities with well-defined properties, such as speed and position. We must consider them as part of a world made up of potentialities and not of objects and facts. The very nature of matter and light becomes subject to interdependent relationships. It is no longer intrinsic, but can change because of an interaction between the observer and the object under observation. Light and matter have no intrinsic reality because they have a dual nature: they appear either as waves or particles depending on the measuring apparatus. The particle and wave aspects cannot be dissociated and complement each other. This is what Bohr called the "principle of complementarily". The phenomenon that we call a "particle" becomes a wave when we are not observing it. But as soon as a measurement is made, it starts looking like a particle again. To speak of a particle's intrinsic reality, or the reality it has when unobserved, would be meaningless because we could never apprehend it. The "atom" concept is merely a convenient picture that helps physicists put together diverse observations of the particle world into a coherent and logical scheme.[4]

Bohr spoke of the impossibility of going beyond the results of experiments and measurements: "In our description of nature, the purpose is not to disclose the real essence of phenomena, but only to track down, so far as possible, relations between the manifold aspects of our experience".[5] Only relationships between objects exist, and not the objects themselves. Quantum mechanics has radically relativised our conception of an object, by making it subordinate to a measurement or, in other words, an event. What is more, quantum fuzziness places a stringent limit on how accurately we can measure reality. There will always be a degree of uncertainty about either the position or the speed of a particle. Matter has lost its substance.

Impermanence at the Heart of Reality

In Buddhism, the concept of interdependence is also closely linked to the notion of the impermanence of phenomena. Buddhism distinguishes two types of impermanence. There is first the gross impermanence – the changing of seasons, the erosion of mountains, the passage from youth to old age, our varying emotions. Then there is the subtle impermanence: at each infinitesimal moment, everything that seems to exist changes. The universe is not made up of solid, distinct entities, but is like a vast stream of events and dynamic currents that are all interconnected and constantly changing. This concept of perpetual, omnipresent change chimes with modern cosmology. Aristotle's immutable heavens and Newton's static universe are no more. Everything is moving, changing and is impermanent, from the tiniest atom to the entire universe, including the galaxies, stars and mankind.

The universe is expanding because of the initial impulse it received from its primordial explosion. This dynamic nature is described by the equations of General Relativity. With the Big Bang theory, the universe has acquired a history. It has a beginning, a past, present and future. It will die in an infernal conflagration or else an icy freeze. All of the universe's structures – planets, stars, galaxies and galaxy clusters – are in perpetual motion and take part in an immense cosmic ballet. They rotate about their axes, orbit, fall toward or move apart from one another. They, too, have a history. They are born, reach maturity, then die. Stars have life cycles that span millions, or even billions of years.

Impermanence also rules the atomic and subatomic world. Particles can modify their nature: a quark can change its family or "flavor", a proton can become a neutron and emit a positron and a neutrino. Matter and antimatter annihilate each other to become pure energy. The energy in the motion of a particle can be transformed into another particle, or vice versa. In other words, an object's property can become an object. Because of the quantum uncertainty of energy, the space around us is filled with an unimaginable number of "virtual" particles, with fleeting ghost-like existences. Constantly appearing and disappearing, they are a perfect illustration of impermanence with their infinitely short life cycles.

Impermanence characterizes also the biological world. Darwin's theory of evolution says that life forms have unceasingly evolved under the influence of natural selection to adapt to environmental changes.

The Anthropic Principle

Despite the remarkable convergences outlined above, there is one area where Buddhism may enter in conflict with modern cosmology. This concerns the anthropic principle that says the universe has had a beginning and that it has been fine-tuned to an extreme degree for the emergence of life and consciousness.

Copernicus's Ghost

Since the sixteenth century, the place of humanity in the universe has shrunk considerably. In 1543, Nicholas Copernicus knocked the earth off its pedestal as the center of the universe by demoting it to the rank of just another planet revolving round the sun. Ever since, the ghost of Copernicus has not ceased to haunt us. If our planet wasn't at the center of the universe, then, our ancestors thought, the sun must be. Then it was discovered that it is just a suburban star among the hundreds of billions of stars that make up our galaxy. We now know that the Milky Way is only one of the several hundred billions of galaxies in the observable universe, which has a radius of about fifteen billion light-years. Humanity is just a grain of sand on the vast cosmic beach. The shrinking of our place in the world led to French philosopher Blaise Pascal's cry of despair in the seventeenth century: "The eternal silence of endless space terrifies me".[6] His anguish was echoed three centuries later by the French biologist Jacques Monod in his book *Chance and Necessity:* "Man knows at last he is alone in the unfeeling immensity of the universe, out of which he has emerged only by chance",[7] and by physicist Steven Weinberg, who remarked: "The more the universe seems comprehensible, the more it also seems pointless".[8]

A Fine-tuning of Unimaginable Precision

I do not think that human life and consciousness arose purely by chance in an unfeeling universe. To my mind, if the universe is so large, then it evolved this way to allow us to be here. Modern cosmology has discovered that the conditions that allow for an intelligence to emerge seem to be coded into the properties of each atom, star and galaxy in our universe and in all of the physical laws that govern it. The universe appears to have been very finely tuned in order to produce a set of intelligent observers capable of appreciating its organization and harmony. This statement is the basis of the "anthropic principle", from the Greek *anthropos* meaning "person". There are two remarks to be made. First, the term "anthropic" is really inappropriate, as it implies that humanity in particular was the goal toward which the universe has evolved. In fact, anthropic arguments would apply to any form of intelligence in the universe. Second, the definition I gave above concerns only the "strong" version of the anthropic principle. There is also a "weak" version that doesn't presuppose any intention in the design of nature. It almost comes down to a tautology – the properties of the universe must be compatible with the existence of humankind – and I will not discuss it further.

What is the scientific basis of the anthropic principle? The way our universe evolved depended on two types of information: (1) its initial conditions, such as its total mass and energy content, its initial rate of expansion, etc. and (2) about fifteen physical constants: the gravitational constant, the Planck constant, the masses of the elementary particles, the speed of light, etc.[9] We can measure the values of these constants with extreme precision, but do not have any theory to predict them. By constructing "model universes" with varying different initial conditions and physical constants, astrophysicists have discovered that these need to be fine-tuned to the extreme: if the physical constants and the initial conditions were just slightly different, we wouldn't be here to talk about them.

For instance, let's consider the initial density of matter in the universe. Matter has a gravitational pull that counteracts the force of expansion from the Big Bang and slows down the universe's rate of expansion. If the initial density had been too high, then the universe would have collapsed into itself after some relatively short time – a million years, a century or even just a year, depending on the exact density. Such a time span would have been too short for stars to accomplish their nuclear alchemy and produce heavy elements like carbon, which are essential to life. On the other hand, if the initial density of matter had been too low, then there would not have been enough gravity to provoke the gravitational collapse of gas clouds to form stars. And if there are no stars, there would be no heavy elements. Life would be impossible as we are all stardust. Everything hangs on an extremely delicate balance. The initial density of the universe had to be fixed to an accuracy of 10^{-60}. This astonishing precision is analogous to the dexterity of an archer hitting a one-centimetre-square target placed fifteen billion light-years away, at the edge of the observable universe! The precision of the fine-tuning varies, depending on the particular constant or initial condition, but in each case, just a tiny change makes the universe barren, devoid of life and consciousness.

Chance or Necessity?

How to account for that extraordinary fine-tuning? It seems to me that we are faced with two distinct choices: the tuning was the consequence of either chance or necessity (to quote the title of Monod's book). If we opt for chance, then we must postulate an infinite number of other parallel universes in addition to our own (these multiple universes form what is sometimes called a multiverse). Each of these universes will have its own combination of physical constants and initial conditions. By chance, ours was the only universe born with just the right combination to create life. All the others were losers and ours is the only winner. If you play the lottery an infinite number of times, then you inevitably end up winning the jackpot. On the other hand, if we reject the hypothesis of a multiverse and adopt the hypothesis of a single universe, ours, then we must postulate the existence of a principle of creation that finely adjusted the evolution of the universe at its very beginning.

How to decide between chance and necessity? Science cannot help us to choose between these two options. In fact, there are several different scientific scenarios that allow for multiple universes. For example, Hugh Everett has proposed, to get around the probabilistic description of the world by quantum mechanics, that the universe splits into as many nearly identical copies of itself as there are possibilities and choices to be made. Some universes would differ by only the position of one electron in one atom, but others would be more radically different. Their physical constants, initial conditions and physical laws wouldn't be the same. Another scenario is that of a cyclical universe with an infinite series of Big Bangs and Big Crunches. Whenever the universe is reborn from its ashes to begin again in a new Big Bang, it would start with a new combination of physical constants and initial conditions. A third possibility is the theory proposed by Andrei Linde whereby each of the infinite number of fluctuations of the primordial quantum froth created a universe. Our universe would then be just a tiny bubble in a super-universe made up of an infinite number of other bubbles. None of those universes would have intelligent life, because their physical constants and laws wouldn't be suitable.

Intriguing as these notions are, I do not subscribe to the idea of multiple universes. The fact that all of these many universes would be unobservable, and thus unverifiable, contradicts my view of science. Science becomes metaphysics when it is no more subjected to the test of experimental proof. Furthermore, Ockham's razor bids us to cut out all the hypotheses that are not necessary: why create an infinite number of barren universes just to produce one that is conscious of its own existence?

In my work as an astronomer, I often have the good luck to travel to observatories to contemplate the cosmos. I am always awed by its organization, beauty, harmony and unity. It is hard for me to think that all that splendor is but the product of pure chance. If we reject the idea of multiple universes and postulate the existence of just one universe, ours, then it seems to me that, we must wager, just like Pascal, on the existence of a creative principle responsible for the fine-tuning of the universe. For me, this principle is not a personified god. It is rather a pantheistic principle that is omnipresent in Nature, not unlike the one described by Einstein and Spinoza. Einstein puts it like this: "The scientist is possessed by the sense of universal causation..."[10] His religious feeling takes the form of a rapturous amazement at the harmony of natural law, which reveals an intelligence of such superiority that, compared with it, all the systematic thinking and acting of human beings is an utterly insignificant reflection. He added: "I believe in Spinoza's God who reveals himself in the harmony of all that exists, but not in a God who concerns himself with the fate and actions of human beings".[11]

Buddhism Denies the Existence of a Creative Principle

The Pascalian wager I just outlined is contrary to the Buddhist approach, which denies the existence of a creative principle (or a watchmaker God). The existence of an entity called God would violate Buddhism's basic tenet of interdependence as that entity can exist by itself, independently of everything else. Buddhism considers that the universe doesn't need any fine-tuning for consciousness to exist. According to it, both have always coexisted, so they cannot exclude each other. Their mutual suitability and interdependence is the precondition for their coexistence.

I am not totally at ease with this explanation. While I admit that this might explain the fine-tuning of the universe, it seems far less clear to me that Buddhism can answer existential questions, of the sort that Leibniz asked about the universe: "Why is there something, rather than nothing?" I would add: "Why are the natural laws as they are and not different?" For example, it would be quite easy to imagine us living in a universe governed only by Newton's laws. But this isn't the case. The laws of quantum mechanics and relativity rule the known universe. What is the origin of these laws?

The Buddhist view also raises other questions. If there is no creator, the universe cannot have been created. So there is neither a beginning nor an end. The only sort of universe that would be compatible with this idea is a cyclical one, with an endless series of Big Bangs and Big Crunches. But the scenario of the universe one day collapsing into itself in a Big Crunch is far from being proven scientifically. It all depends on the amount of dark matter and dark energy in the universe, and this is as yet unknown.

According to the latest astronomical observations, the universe does not seem to have enough dark matter to stop and reverse its expansion. They seem to indicate that we live in a flat universe that will expand forever and will stop only after an infinite time. Thus our present state of knowledge seems to exclude the idea of a cyclical universe.

Buddhism postulates also that there exists streams of consciousness that have coexisted with the material universe since the first fractions of a second after the Big Bang. Science is still far from being able to examine experimentally this postulate. Some neurobiologists think that there is no need for consciousness streams that coexist with matter, and that maybe consciousness can arise from matter by emergent processes, once matter in its path towards more organization has passed a certain threshold of complexity.

Science and Spirituality: Two Windows into Reality

I have attempted to show that there are striking convergences between the views of reality from modern science and from Buddhism. The concept of interdependence, which is at the heart of Buddhism, is echoed by the globality of reality implied by the EPR experiment on the subatomic and atomic scale, and by Foucault's pendulum and Mach's principle on the scale of the universe. The

Buddhist concept of "emptiness", the absence of intrinsic existence, finds its scientific equivalent in the dual nature of light and matter in quantum mechanics. Because a photon is a wave or a particle depending on how we observe it, it cannot be said to exist as an entity with an inherent existence. The concept of impermanence echoes the concept of evolution in cosmology and biology. Nothing is static, everything changes, moves and evolves, from the tiniest atom to the largest structures in the universe. The universe itself has acquired a history. Life itself has had a long history before the emergence of *Homo sapiens*.

I have also pointed out a potential area of disagreement: Buddhism rejects the idea of a beginning of the universe and of a creative principle (or creator) that fine-tunes its properties for the emergence of life and consciousness.

The above convergences are not surprising, since both science and Buddhism use criteria of rigor and authenticity to attain the truth. Since both aim to describe reality, they must meet on common grounds and not be exclusive of each other. Whereas in science the primary methods of discovery are experimentation and theorizing based on analysis, in Buddhism contemplation is the primary method. Both are windows that allow us to peer at reality. They are both valid in their respective domains and complement each other.

Science reveals to us "conventional" knowledge. Its aim is to understand the world of phenomena. Its technical applications can have a good or bad effect on our physical existence. Contemplation, however, aims to improve our inner selves so that we can improve everybody's existence. Scientists use ever more powerful instruments to probe Nature. In the contemplative approach, the only instrument is the mind. The contemplator observes how his thoughts are bound together and how they bind him. He examines the mechanisms of happiness and suffering and tries to discover the mental processes that increase his inner peace and make him more open to others in order to develop them, as well as those processes that have a destructive effect in order to eliminate them. Science provides us with information, but brings about no spiritual growth or transformation. By contrast, the spiritual or contemplative approach must lead to a profound personal transformation in the way we perceive the world and act on it. The Buddhist, by realizing that objects have no intrinsic existence, lessens his attachment to them, which diminishes his suffering. The scientist, with the same realization, is content to consider that as an intellectual advance which can be used to advance his work, without changing fundamentally his basic vision of the world and how he leads his life.

When faced with ethical or moral problems that, as in genetics, are becoming ever more pressing, science needs the help of spirituality in order not to forget our humanity. As Einstein puts it so well:

> The religion of the future will be a cosmic religion. It will have to transcend a personal God and avoid dogma and theology. Encompassing both the natural and the spiritual, it will have to be based on a religious sense arising from the experience of all things, natural and spiritual, considered as a meaningful unity… Buddhism answers this description… If there is any religion that could respond to the needs of modern science, it would be Buddhism.[12]

NOTES

[1] M. Ricard and Trinh Thuan, *The Quantum and the Lotus: A Journey to the Frontiers where Science and Buddhism Meet* (New York: Crown, 2001).
[2] W. Heisenberg, *Physics and Beyond: Encounters and Conversations* (New York: Harper and Row, 1958).
[3] A. Einstein, B. Podolsky and N. Rosen, "Can Quantum-Mechanical Description of Physical Reality be Considered Complete?" *Physical Review,* 1935, Vol. 41, p. 777.
[4] P. C. W. Davies and J. R. Barnes, eds., *The Ghost in the Atom* (Cambridge: Cambridge University Press, 1993).
[5] N. Bohr, *Atomic Theory and the Description of Nature* (Cambridge: Cambridge University Press, 1934).
[6] Blaise Pascal, *Pensees* (London: Penguin Classics, 1660).
[7] Jacques Monod, *Chance and Necessity* (New York: Knopf, 1971).
[8] Steven Weinberg, *The First Three Minutes* (New York: Basic Books, 1977).
[9] J. Barrow and F. Tipler, *The Cosmological Anthropic Principle* (Oxford: Oxford University Press, 1986).
[10] Alice Calaprice, ea., *The Quotable Einstein* (Princeton: Princeton University Press, 1996), p. 151.
[11] *Ibid.,* p. 147.
[12.] *Ibid.,* p. 161.

Part 2. Contributed Papers

Section A. Stars and Stellar Remnants

International Symposium on Astrophysics Research and on the Dialogue between Science and Religion, Vatican Observatory, 2003
C. D. Impey & C. E. Petry, editors

Black Hole and Brane Factories

Eun-Joo Ahn
Department of Astronomy and Astrophysics, University of Chicago, 5640 S. Ellis Avenue, Chicago, IL, 60637, USA

Abstract. An alternative way of explaining the hierarchy problem is to lower the gravity scale to the TeV scale. In this scenario, scattering of particles with center-of-mass energy of the order of a few TeV can lead to the creation of nonperturbative, higher-dimensional gravitational objects: black holes and branes. Low-energy branes could dramatically change our picture of high-energy physics. Black holes and branes may be created in future particle colliders, observed from ultra high energy cosmic rays, and constitute dark matter.

1. Introduction

We observe two different fundamental energy scales in nature: the electroweak scale, $E_{EW} \sim 1$ TeV, and the gravitational scale, $E_G \sim 10^{16}$ TeV. Unification of the two scales must be encoded in any Grand Unification Theory: either the gravitational scale is lowered to the electroweak scale, or the electroweak scale is brought up to the gravitational scale. It has been proposed that E_G may be brought down to $\sim$ TeV with presence of large[1] extra dimensions (LEDs) (Antoniadis 1990, Arkani-Hamed et al. 1998, and Antoniadis et al. 1998). Only gravity can see these LED and thus the weakness of gravity can be explained.

A striking consequence of LEDs is the creation of non-perturbative gravitational objects (NPGOs), such as black holes and p-branes (Banks & Fischler 1999, Ahn et al. 2003). A series of papers in recent years (see Cavaglià 2003 for a comprehensive review and complete references) has proposed black hole (Giddings & Thomas 2002, Dimopoulos & Landsberg 2001) and p-brane formation in particle colliders and via cosmic rays in the atmosphere (Ahn et al. 2003, Anchordoqui et al. 2002, Feng & Shapere 2002). Decay signature of these NPGOs will be different from other standard model phenomena. For the first time, quantum gravity stands a chance of being experimentally tested. In this paper we give a brief overview of the important consequences of the TeV gravity scenario.

[1] "Large" in this context means larger than the fundamental Planck scale.

2. TeV Gravity

The 4-dimensional Planck mass in natural units is given by $M_{Pl} \equiv G_4^{-1/2}$, where G_4 is the 4-dimensional gravitational constant (Newton's constant). When extra dimensions are considered the fundamental Planck scale is given by $M_\star = G_{n+4}^{-1/(n+2)}$, where n is the number of extra dimensions. The 4-dimensional Newton's constant and the $(n+4)$-dimensional gravitational constant are related by

$$G_4 = \frac{G_{n+4}}{V_n} \;\rightarrow\; M_{Pl}^2 = M_\star^{n+2}\, V_n\,, \tag{1}$$

where V_n is the volume of the extra dimensions. An example of a simple compactification is the toroidal compactification with two length scales L and L' (Lykken & Nandi 2000). The relation between M_{pl} and $M_\star$ is given by

$$\left(\frac{M_{Pl}}{M_\star}\right)^2 = \left(\frac{L}{L_\star}\right)^m \left(\frac{L'}{L_\star}\right)^{n-m}, \tag{2}$$

where there are m dimensions of L and $(n-m)$ dimensions of L' ($m \leq n$), and $L_\star = M_\star^{-1}$. Newton's Law is known to hold until $\sim 150\ \mu$m (Adelberger 2002). The case of $n = 1$ where $L \sim 10^{13}$ cm is ruled out. However $n = 2$ gives $L \sim 100\,\mu$m which is still possible.

3. Black Hole and p-Brane Formation

NPGO formation is a generic non-perturbative phenomenon of any gravitational theory: any process with centre-of-mass larger than the Planck scale concentrated within the event horizon can generate a black hole. The (n+4)-dimensional Einstein-Hilbert action is given by

$$S_{EH} = \frac{M_\star^{n+2}}{16\pi} \int d^{n+4} \sqrt{-g}\, \mathcal{R}(g)\,, \tag{3}$$

where g is the determinant of the metric $g_{\mu\nu}$, and $\mathcal{R}(g)$ is the Ricci scalar. A static, neutral p-brane is described by (Gregory 1996, Cavaglià 1997) as

$$ds^2 = R^{\frac{\Delta}{p+1}}(-dt^2 + dz_i^2) + R^{\frac{2-q-\Delta}{q-1}} dr^2 + r^2 R^{\frac{1-\Delta}{q-1}} d\Omega_q^2\,, \tag{4}$$

where

$$R(r) = 1 - \left(\frac{r_p}{r}\right)^{q-1}, \; \Delta = \sqrt{\frac{q(p+1)}{p+q}}\,. \tag{5}$$

One recovers the Schwarzschild metric for the case of $p = 0$. The cross section for the brane is given as

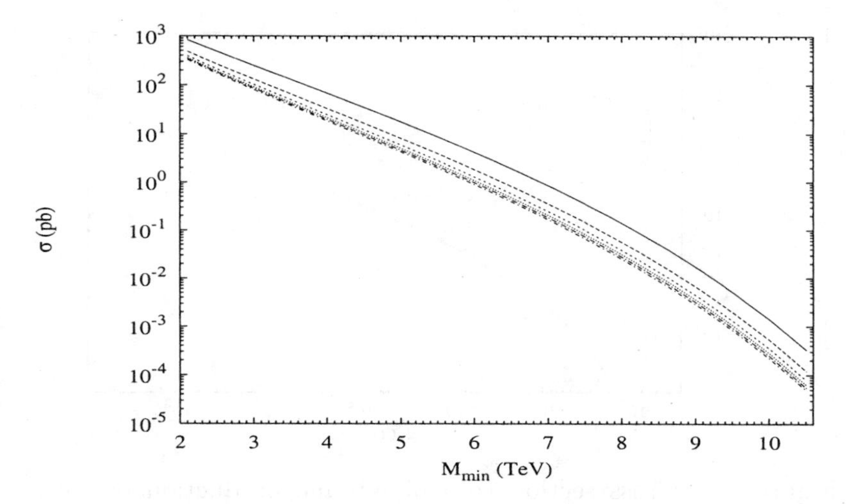

Figure 1. Cross sections (pb) of a black hole formation ($n = 2$ to $n = 7$ from above) by proton-proton scattering at the LHC. $M_\star$ is assumed to be 1 TeV.

$$\sigma_{ij \to br}(s; n, p, V_p) \approx \pi r_p^2 = \frac{1}{M_\star^2} \gamma(n,p)^2 \, V_p^{-\frac{2w}{n+1}} \left(\frac{M}{M_\star}\right)^{\frac{2w}{n+1}} , \qquad (6)$$

where V_p is the volume of the extra dimensions in fundamental Planck units where the p-brane wraps, $w = [1 - p/(n+1)]^{-1}$, and $\gamma(n,p)$ is a factor of order $O(1)$. Cross section for black holes can be obtained with $p = 0$.

4. Black Hole and Brane Factories

The Large Hadron Collider, a proton-proton collider with centre-of-mass (CM) energy of 14 TeV, is currently being built at CERN. With a luminosity of $\sim 3 \times 10^5\,\mathrm{pb}^{-1}\mathrm{yr}^{-1}$, BHs and branes can be created at a rate of approximately 1 per second. The cross section, hence the formation rate, depends on the minimum mass at which the NPGO forms (M_{min}). The cross section is plotted in Figure 1. NPGOs will be distinguishable from the standard model process due to different decay signatures.

Ultra high-energy cosmic rays are another way to test TeV gravity. Interaction between cosmic rays with energy $\geq 5 \times 10^{19}$ eV and blueshifted (in cosmic ray's frame) cosmic microwave background radiation creates high energy neutrinos. These neutrinos have high enough center of mass energy when they collide with particles in the Earth's atmosphere and are able to create NPGOs. Interactions where neutrinos enter the Earth's atmosphere horizontally, giving ~ 40 times more interaction length than a

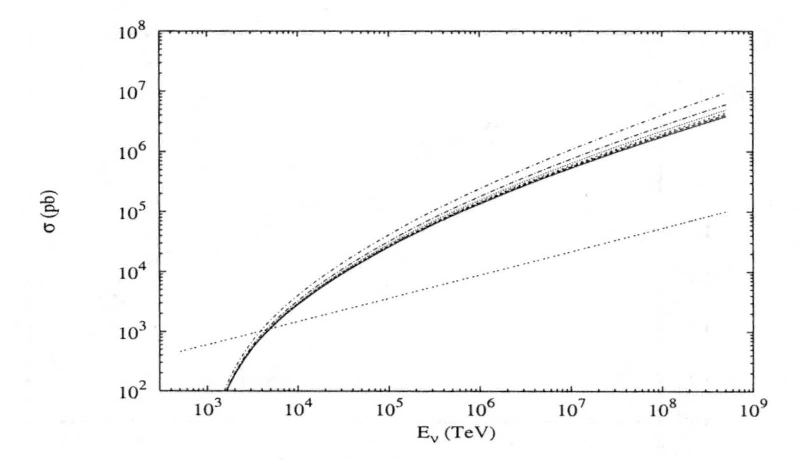

Figure 2. Cross section (pb) of p-brane production by ultra high energy cosmic rays for $n = 7$ ($p = 0$ to $p = 5$ from below). $M_\star = M_{min} = 1$ TeV has been assumed. The dashed curve is for the standard model process.

vertically incoming particle, are most likely to occur. Cosmic ray detectors such as the Pierre Auger Observatory will be able to observe these in the form of deeply penetrating horizontal air showers. Cross sections of NPGO formation are substantially higher than the standard model process (see Fig. 2), giving a higher formation rate. Different air shower characteristics and statistics will indicate the formation of NPGOs.

5. Discussion and Conclusion

NPGO formation could play a fundamental role in all physical processes with energy above the TeV scale. In particle accelerators and cosmic ray physics, formation and decay of NPGOs are expected to be instantaneous ($\sim 10^{-26}$ sec) and distinguishable from other standard model processes. Some subset of supersymmetric p-branes are known to be stable; they could be dark matter candidate as the early universe is an ideal place for them to form (Ahn & Cavaglià 2002). If the LED scenario is the correct description of the world, theories of quantum gravity may be experimentally testable for the first time.

Acknowledgments. I am grateful to my collaborators Marco Cavaglià and Angela Olinto for carrying out this research together. I thank Father Coyne for giving both moral and material support and leaving such beautiful memories of Castel Gandolfo and Rocca di Papa.

References

Adelberger, E.G. (EOT-WASH Group Collaboration) 2002, hep-ex/0202008
Ahn, E.J., Cavaglià, M. & Olinto, A.V. 2003, Phys. Lett. B, 551, 1
Ahn, E.J. & Cavaglià, M. 2002, Gen. Rel. Grav., in press, hep-ph/0205168
Anchordoqui, L., Feng, J.L., Goldberg, H. & Shapere, A.D. 2002, Phys. Rev. B, 65, 124027
Antoniadis, I. 1990, Phys. Lett. B, 246, 377
Antoniadis, I., Arkani-Hamed, N., Dimopoulos, S. & Dvali, G.R. 1998, Phys. Lett. B., 436, 257
Arkani-Hamed, N., Dimopoulos, S. & Dvali, G.R. 1998, Phys. Lett. B., 429, 263
Banks, T. & Fischler, W., 1999, hep-th/9906038
Cavaglià, M. 1997, Phys. Lett. B., 413, 287
Cavaglià, M. 2003, Int. Jou. Mod. Phys. A, in press, hep-ph/0210296
Dimopoulos, S. & Landsberg, G. 2001, Phys. Rev. Lett., 87, 161602
Feng, J.L. & Shapere, A.D. 2002, Phys. Rev. Lett., 88, 021303
Giddings, S. & Thomas, T. 2002, Phys. Rev. D, 65, 056010
Gregory, R. 1996, Nucl. Phys. B, 467, 159
Lykken, J. & Nandi, S. 2000, Phys. Lett. B, 485, 224

International Symposium on Astrophysics Research and on the Dialogue between Science and Religion, Vatican Observatory, 2003
C. D. Impey & C. E. Petry, editors

Cosmic Rays, Stellar Evolution and Supernova Physics

Peter L. Biermann[1,2]

[1] *Max-Planck Institut für Radioastronomie, D-53121 Bonn, Germany*
[2] *Department for Physics and Astronomy, University of Bonn*
plbiermann@mpifr-bonn.mpg.de

Ana Vasile[1,3]

[3]*Institute of Space Sciences, Bucharest-Magurele, Romania*
avasile@venus.nipne.ro

Abstract. For some time it has been thought that the explosions of massive stars produce cosmic rays, in a kind of relativistic tennis game across the shock. The model has been made more specific by arguing about which kind of star provides what kind of cosmic rays, e.g., nuclei of which chemical elements, what spectrum, and what kind of interaction with nuclei in the interstellar medium. The recent success of a quantitative proposal from ten years ago now allows some rather strong conclusions to be drawn: (1) Massive stars beyond about 25 solar masses at zero age all converge to a common final state; they all become very similar to each other near of the end of their lifetime; (2) the supernova mechanism must involve magnetic fields and rotation, just as suggested by Bisnovatyi-Kogan more than 30 years ago; (3) the explosion produces about 10^{52} erg in explosive energy, rather similar to what has been suggested for gamma ray bursts; (4) very likely such explosions may serve as a standard candle in cosmological work, if we can work out the correction for the extreme anisotropy of such explosions.

1. Introduction

Several seemingly unrelated questions appear in high energy work, and in the search for an understanding of stellar evolution, supernovae, and gamma ray bursts:

- Where do cosmic rays come from? Where is the transition between Galactic cosmic rays and extragalactic cosmic rays?

- What is the physics of supernova explosions? What are the final states of stars just before they blow up? How does this physics change as the mass of the star goes up? Are these explosions isotropic?
- What is the relation to gamma ray bursts? Are gamma ray bursts just a special class of supernovae, or are they a successor to an earlier supernova, or are they completely unrelated?
- Do very massive stars provide another, maybe even brighter, standard candle useful in cosmology? Under what conditions might this be possible?

After cosmic rays were discovered by Heß and Kohlhörster just after 1910, it had been suggested by Baade and Zwicky already in the 1930s that cosmic rays get energized by supernova explosions. How this happens was clarified by first by Fermi in 1949, and by many others in later work. The process can be described as a relativistic tennis game, where the flow on two sides of a shock wave corresponds to the motion of the tennis racket towards the net at the point of each hit, and magnetic resonant scattering plays the role of the tennis racket. The quantitative theory of this process was developed in several parallel papers by Axford et al., Bell, Krymsky, Blandford and Ostriker in 1977/1978, and tested with Solar wind shocks.

When a supernova explodes, a powerful shock wave travels through its environment, and produces very energetic particles. The spectral data of cosmic rays suggest that this can account for all particles in cosmic rays up to an energy of about 3×10^{18} eV. How this can happen in detail is far from agreed upon.

In a series of papers about ten years ago we suggested in collaboration with many of our friends, that it is possible to develop a quantitative theory to explain spectra, abundances, and quite possibly all details of the observations of cosmic rays, focussing on the fact that many very massive stars have a powerful wind. It follows that cosmic rays may reflect the important properties of this wind, as well as some key properties of the explosion.

2. Cosmic Rays

Cosmic rays show a spectrum, which is a power law, up to a bend in the spectrum, the "knee" at about 3×10^{15} eV, where the spectrum bends down by about 0.4 in spectral index. Near 3×10^{17} eV, the downward trend accelerates, at the "second knee," to hit a new component near 3×10^{18} eV, presumably the extragalactic component rising up. One of the major difficulties has been to find a plausible explanation for this "knee" feature. The proposal made ten years ago was that this reflects the transition from parallel to perpendicular acceleration, in a shock, which travels through a wind, where the magnetic field is a tightly wound spiral; this entails that the shock is perpendicular over most of 4π, and only in a small polar region does the radial field dominate, and so the parallel configuration prevails. In other words, the shock normal is perpendicular to the magnetic field over much of the sphere of a shock in a wind.

The magnetic field configuration was derived by Parker, the parallel acceleration has been treated by Lagage and Cesarsky, and the perpendicular case by Jokipii. Since a simple one-star model fits the cosmic ray spectrum, which is clearly the superposition of many contributing supernovae, there are serious consequences. Such an argument requires that all contributing sources must be the same, and therefore the quantitative values of magnetic field, the shock velocity, as well as the surface rotation of the star just before it blows up, must the same, or at least very nearly the same, for all stars contributing. Such a tight relationship between magnetic field, rotation and explosive energy is a natural consequence of the magnetorotational mechanism proposed by Bisnovatyi-Kogan in 1970, following a broader suggestion by Kardashev in 1964.

The model has been rather successful in interpreting the data, especially from KASCADE, but also under the constraints to fit the higher energy data from Akeno, HiRes, Haverah Park, Yakutsk, etc. So indeed, we conclude that the supernova mechanism is what Bisnovatyi-Kogan suggested. This statement is true only for Wolf-Rayet stars which end their lives as blue supergiants. In fact, it immediately implies that all such stars are very nearly the same at the stage of explosion at the end of their evolution.

We can then fit quantitatively the abundances, matching say, carbon/oxygen to the abundances in the wind of such stars, and the flux in cosmic rays to the energetics of such explosions, and find that the explosive energy is about 10^{52} erg. This has been confirmed by the best observed case, SN 1998bw. It also follows that such explosions are very anisotropic, also confirmed by SN 1998bw.

Interestingly, this energy of explosion corresponds well to the scale of energy expected for gamma ray bursts, which are believed to be a special case of a supernova, perhaps even generally the formation event of a black hole. Fitting the fluence distribution of gamma ray bursts (Figure 1), and allowing for the selection effects, gives an energy scale as well, and this is also 10^{52} erg. In the first treatment this energy emerges as a maximum of a distribution, and not yet as a fixed number.

3. Cosmology

If it is really true that the final configuration of these stars is so very similar just before their explosion, then presuambly all aspects of the explosion are similar, and so they may become a new standard candle in cosmology, possibly a much brighter standard candle than the famous Supernovae of type Ia. There is a catch, however—in the mechanism of Bisnovatyi-Kogan,

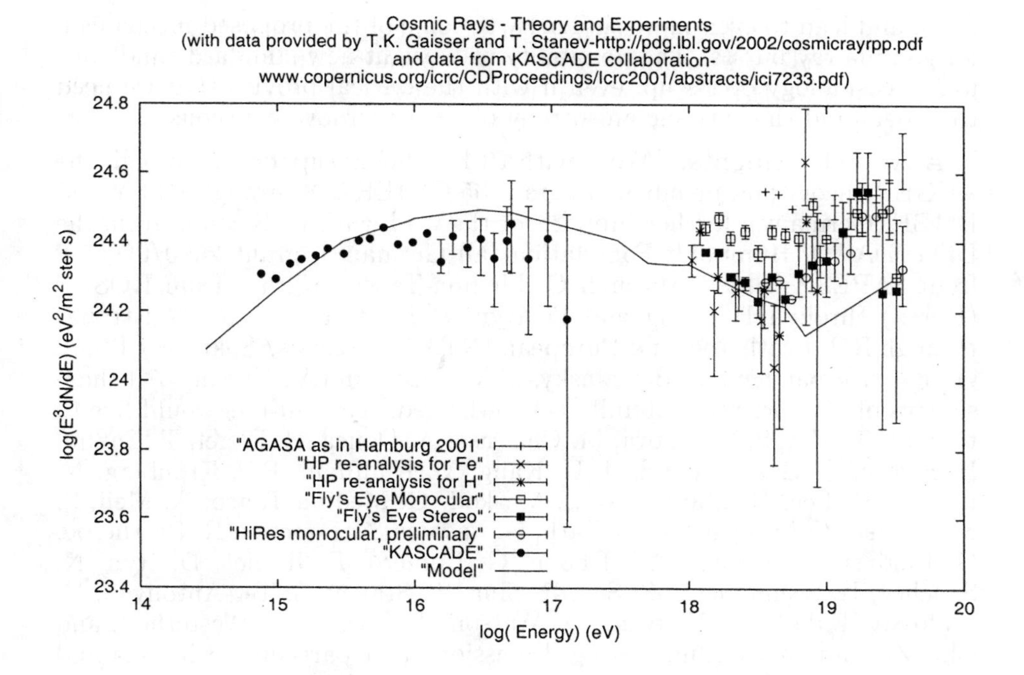

Figure 1. The fluence distribution for cosmic rays, measured by a variety of facilities over the past few years.

the explosion is very anisotropic, and so, with some clever further work we need to first understand how to disentangle this anisotropy before we can safely proceed. Interestingly, this expected anisotropy is rather smilar to what we know already about Active Galactic Nuclei, where rotationally symmetric models with extreme Lorentz beaming along a relativistic jet direction parallel to the axis of the symmetry plane of the activity, are the standard today. If there were such a standard candle, then it could be much brighter than the Supernovae of type Ia, which are so very famous today, and have led to the development of arguments in favor of a cosmological dark energy dominating over dark matter, and certainly over baryonic matter.

4. Conclusion

A proposal for the origin of cosmic rays has led us to some interesting conclusions. The physics of cosmic rays have led us to put some constraints on massive star evolution, to ascertain the physical mechanism of the supernova explosion of the most massive stars, the magneto-rotational mechanism, and lead to even more similarity to some of the proposed mechanism for gamma ray bursts. The possibility of a bright new standard candle for use in cosmology opens up, even if with the critical proviso that we need to understand the extreme anisotropy of the supernova exlosions.

Acknowledgments. Work with PLB is being supported through the AUGER theory and membership grant 05 CU1ERA/3 through DESY and BMBF (Germany); further support for the work with PLB comes from the DFG, DAAD, Humboldt Foundation (all Germany), grant 2000/06695-0 from FAPESP (Brasil) through G. Medina-Tanco, a grant from KOSEF (Korea) through H. Kang and D. Ryu, a grant from ARC (Australia) through R.J. Protheroe, and European INTAS/ Erasmus/ Sokrates/ Phare grants with partners V. Berezinsky, M.V. Rusu, and V. Ureche. All these sources of support are gratefully acknowledged. The authors would like to thank E.J. Ahn, S. Casanova, M. Chirvasa, A. Donea, R. Engel, T. Enßlin, H. Falcke, C. Galea, R. Engel, H. Kang, M. Kaufman, P.P. Kronberg, N. Langer, H. Lee, K. Mannheim, S. Markoff, G. Medina-Tanco, A. Meli, S. Moiseenko, F. Munyaneza, B. Nath, G. Pavalas, A. Popescu, R. Protheroe, G. Pugliese, J. Rachen, W. Rhode, G. Romero, E. Roulet, D. Ryu, N. Sanchez, H. Seemann, E.-S. Seo, R. Sina, T. Stanev, S. Ter-Antonyan, V. Tudose,t H. de Vega, Y. Wang, A. Watson, T. Weiler, St. Westerhoff, and Ch. Zier for many illuminating discussions and partnership in this and related work. More details can be found through the group's web-page www.mpifr-bonn.mpg.de/div/theory/.

International Symposium on Astrophysics Research and on the Dialogue between Science and Religion, Vatican Observatory, 2003
C. D. Impey & C. E. Petry, editors

Spectroscopic Indices of Stars: A Theoretical Approach

Miguel Chavez, Lino H. Rodriguez-Merino and Emanuele Bertone

INAOE-Tonantzintla, Luis Enrique Erro No.1, Tonantzintla, 72840 Puebla, México

Alberto Buzzoni

Telescopio Nazionale Galileo, Santa Cruz de La Palma, Spain and Osservatorio Astronomico di Brera, Milan, Italy

Abstract. We present a new grid of theoretical absorption indices based on an updated library of synthetic spectra. These were computed using a subset of LTE Kurucz model atmospheres and his series of codes SYNTHE. Spectra encompass the wavelength interval from 850 to 7000 Å, with an inverse resolution of $R = 50,000$ (UV range) and $500,000$ (optical). A wide space domain in the atmosphere parameters has been probed such as to account for stars along the whole spectral-type sequence (from O to M) and luminosity class (MK type from I to V). Metallicity ranges from $[Fe/H] = -3 \rightarrow +0.5$ dex. A total of 51 ultraviolet and optical narrow-band indices defined in the literature have been measured and calibrated in terms of the atmosphere parameters. We foresee extensive use of this grid for stellar and extragalactic studies.

1. Introduction

Over the past decade the study of spectral properties of stars, through the definition of narrow-band indices, has received considerable attention. Spectroscopic indices have quickly become a basic and popular interpretative tool, not only for the study of single stars (e.g. Gray, Scott & Postma 2002), but also to assess the overall evolutionary status of stellar aggregates as a whole (e.g. Buzzoni, Gariboldi & Mantegazza 1992; Worthey et al. 1994). There is at least a threefold advantage in using such a narrow-band approach to stellar spectra:

1. Indices can easily be measured from stellar or galaxy spectral energy distributions.
2. They do not depend on interstellar extinction.
3. Different atomic or molecular species can be studied separately.

In addition, narrow-band indices are especially suitable to analytical handling via the so-called fitting functions (i.e. polynomials describing the relation of index behaviour vs. leading atmosphere parameters), thus easing their application for population synthesis purposes (Buzzoni et al. 2001).

Most of the index datasets currently available in the literature rely on empirical spectral libraries. While such databases have played a key role in studying e.g. chemical evolution of galaxies, there are still some important problems to be tackled. First, empirical databases are constructed upon stars in the solar neighborhood, and therefore implicitly carry on the imprints of the local chemical history. Furthermore, observed spectra may usually not be systematically labeled in terms of atmosphere fundamental parameters, and in most libraries the parameter space is not homogeneously covered, especially in the high and very low metallicity regimes (see, for example, Cenarro et al. 2002 for a discussion on the CaII triplet).

This is not the case for theoretical databases, which naturally overcome all these problems (see Chavez, Malagnini & Morossi 1996 and references therein). A theoretical spectrum is in fact labeled by the parent model, and it can be computed for virtually any combination of atmosphere parameters and at any desired wavelength resolution. The material presented in this paper complements and updates previous investigations on the analysis of theoretical spectroscopic indices based on high-resolution synthetic spectra.

2. Model Atmospheres and Library of Synthetic Spectra

For the calculation of theoretical spectra we have extracted a subset of the grid of LTE model atmospheres developed by Kurucz (1993). In brief, these models consider local thermodynamic equilibrium and plane parallel geometry while line opacity is accounted for through a set of opacity distribution functions. The selected subset of models consists of 2482 entries and covers a large volume of the parameter space; effective temperature ($T_{\rm eff}$) in the interval 3500–50,000 K with a variable step ranging from 250 K for models with $T_{\rm eff} \leq 7000$ K to 2500 K for hotter models. Surface gravities vary from 0.0 to 5.0 dex and chemical compositions are $[M/H]$ = –3.0, –2.0, –1.5, –1.0, –0.5, –0.3, 0.0, +0.3, and +0.5. In all cases we have adopted a microturbulence velocity of 2 km s^{-1}.

The library consists of two datasets computed independently, covering two overlapping spectral regions. The first set includes the far ultraviolet to optical region (850–4750 Å) at an inverse resolution of $R = 50,000$. The latter spans from 3500 to 7000 Å at a resolution of $R = 500,000$. Calculations of both sets were carried out by using the SYNTHE series of codes developed by Kurucz (1993) and included the opacities of several million atomic and molecular transitions for the most important diatomic species. It is important to note that unlike previous investigations in the optical region conducted by some of us (Chavez, Malagnini & Morossi

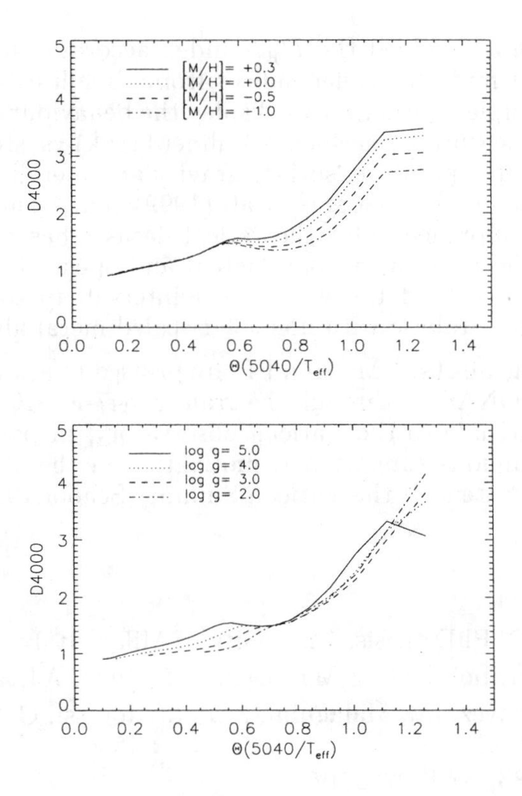

Figure 1. Theoretical D_{4000} index from our Far-UV optical database as a function of the effective temperature parameter, $\Theta_{\rm eff} = 5040/T_{\rm eff}$, vs. chemical composition for dwarf models ($\log g = 5.0$, *upper panel*), and vs. stellar gravity for fixed solar metallicity (*lower panel*). The index gives a measure of the Balmer spectral break around 4000 Å on a magnitude scale.

1996), we have included the opacity due to TiO (Schwenke 1998) and we considered the so-called approximate overshooting in the treatment of convection (Castelli, Gratton & Kurucz, 1997). This composite library largely supercedes any other dataset available in the literature computed, and provides increased resolution. Details of the library and the library itself will be available soon (Bertone 2002; Rodriguez-Merino et al. 2003, in preparation; Bertone et al. 2003, in preparation).

3. The New Grid of Theoretical Indices

Calculation of the 25 spectral indices in the optical range (4080–6400 Å) has been carried out following the Lick/IDS system definitions (Worthey et al. 1994); we also added the D_{4000} index according to Bruzual (1983), and the far and mid ultraviolet indices from Fanelli et al. (1992). As an illustrative example, in Figure 1 we display the behaviour of the D_{4000} index (which gives a measure of the 4000 Å Balmer break) vs. stellar fundamental parameters (i.e. temperature, surface gravity and metallicity). Our results nicely match the work of Gorgas et al. (1999) that, contrarily, relies on a fully empirical approach. Our theoretical dataset has been calibrated as a function of the atmosphere parameters and spectral resolution. Work is in progress to extend the wavelength interval up to 9000 Å, and to complement our database with non solar-scaled metal abundances.

Acknowledgments. MC and EB are pleased to acknowledge financial support from CONACyT through the grant 36547-E. MC wishes to express his deepest gratitude to the Vatican observatory, in particular to Father Coyne for continuous support and motivation in the 12 years that have passed since he attended the Vatican Summer School 1990.

References

Bertone E. 2002, PhD thesis, University of Milan, Italy

Buzzoni, A., Gariboldi, G. & Mantegazza, L. 1992, AJ, 103, 1814

Buzzoni, A., Chavez, M., Malagnini, M.L. & Morossi, C. 2001, PASP, 113, 1365

Bruzual, G. 1983, ApJ, 273, 105

Castelli, F., Gratton, R. & Kurucz, R.L. 1997, A&A, 318, 841

Chavez, M., Malagnini, M. & Morossi, C. 1996, ApJ, 471, 726

Cenarro, A.J., Gorgas, J., Cardiel, N., Vazdekis, A. & Peletier, R.F. 2002, MNRAS, 329, 863

Fanelli, M.N., O'Connell, R.W., Burstein, D. & Wu, C. 1992, ApJS, 82, 197

Gorgas, J., Cardiel, N., Pedraz, S. & Gonzalez, J.J. 1999, A&AS, 139, 29

Gray, D.F., Scott, H.R. & Postma, J.E. 2002, PASP, 114, 536

Kurucz, R.L. 1993, CD-ROM 13, ATLAS9 Stellar Atmosphere Programs and 2 km s^{-1} grid (SAO: Cambridge); CD-ROM 18, SYNTHE Spectrum Synthesis Programs and Line Data (SAO: Cambridge)

Schwenke, D.W. 1998, Faraday Discuss, 109, 321

Worthey, G., Faber, S., Gonzalez, J.J. & Burstein, D. 1994, ApJS, 94, 687

International Symposium on Astrophysics Research and on the Dialogue between Science and Religion, Vatican Observatory, 2003
C. D. Impey & C. E. Petry, editors

Non-Linear Radial Oscillation and Chaotic Behavior of One-Zone Stellar Models

M. Devasagayam, S.J.

St. Joseph's College, Tiruchirapalli 620 002, India

Abstract. We have studied the non-linear oscillations of a simple stellar model based on Baker's One Zone model. It has been shown by Saitou that when the perturbations for radius, pressure and saturation of the excitation mechanism are taken into account, the non-linear oscillator model bifurcates and develops chaotic trajectories through period doubling. In our case, when perturbations of radius and temperature are taken into account, the model develops into a chaotic behavior for a range of control parameter $a = -11$ to $a = -24$.

1. Introduction

Study of stellar pulsations has greatly improved our understanding of stellar structure and evolution. The theory of non-linear dynamical systems has been employed recently to describe pulsation properties of irregular variables like RV Tauri, yellow and red super giants, and ZZ Ceti stars (Buchler and Regev 1982, Perdang and Blacher 1982, Auvergne and Baglin 1985, Perdang 1985, Tanaka and Takeuti 1988, Devasagayam et al. 1999).

Buchler and Regev (1982) and Auvergne and Baglin (1985) have studied models under the assumption that the non-linearity arising due to variations in adiabatic exponent is the main agent for non-linear pulsation. Tanaka and Takeuti (1988) pointed out that a dynamical stability may be necessary for actual models of pulsating stars and modified the nonadiabatic term in an artificial way which he derived from the Rossler (1976) oscillator. Their model succeeded in showing a switch from periodic oscillating systems to quasi periodic ones with the changes of the control parameters. The control parameter usually means the degree of vibrational instability.

The idea of using one-zone models for studying non-linear behaviour of stellar pulsation was put forward by Baker (1966). The elementary one-zone models have suggested the existence of Feigenbaum period doubling bifurcation in stars (Buchler and Regev 1982). On the other hand, Moore and Spiegel (1966) oscillators succeeded in showing irregular oscillations; their equilibrium state is dynamically unstable and seems to be unrealistic for variable stars. So compared to the Moore and Spiegel-type oscillator, the Tanaka and Takeuti (1988) oscillator has an advantage as their model is dynamically stable. However, the assumed non-linear cooling term was not derived from reasonable analyses of the physical processes in stellar outer layers.

We develop another model for the stellar pulsation, which is constructed in a way which is simpler than the previous attractors and also produces a sequence of period-doubling bifurcation. In this model we introduce a small perturbation in temperature following the variations in radius or material displacement along the radius. The model is an extension of the oscillations of the Baker (1966) one-zone stellar model to a finite-amplitude one. We qualitatively compare the first return map of our oscillator with those of several variable stars.

2. Non-Linear Radial Oscillations of the One-Zone Model

The stellar model formulated by Baker (1966) is used here, where a one-zone model consists of a rigid core of constant radius R_c and luminosity L_c surrounded by a simplified envelope. It is assumed that the core radius R_c is small compared with the time-dependent stellar radius R, i.e. that $R_c \ll R$.

Applying the equation of motion to our model, we obtain

$$\frac{d^2R}{dt^2} = \frac{4\pi R^2 P}{m} - \frac{GM}{R^2}, \tag{1}$$

where M is the stellar mass, P the pressure in the shell. The opacity law of Rosseland is given by

$$\kappa = \kappa_0 \rho^n T^{-s}. \tag{2}$$

We define the relative variations of the radius and temperature, x and z as follows:

$$R = R_0(1+x), \tag{3}$$
$$T = T_0(1+z). \tag{4}$$

Accordingly, we get

$$\rho = \rho_0(1+x)^{-3}. \tag{5}$$

Putting,

$$\frac{dx}{dt} = y, \tag{6}$$

we then have

$$\frac{dy}{dt} = \Omega^2\left[(1+x)^{-1}(1+z) - (1+x)^{-2}\right], \tag{7}$$
$$\frac{dz}{dt} = -3y\gamma(1+x)^{-1}(1+z) - \epsilon\left[(1+x)^{\alpha}(1+z)^{\beta} - 1\right], \tag{8}$$

where

$$\Omega^2 = \frac{GM}{R_0^3}, \tag{9}$$

$$\epsilon = \frac{L_0}{mC_vT_0}, \tag{10}$$

$$\alpha = 4 + 3n, \tag{11}$$

$$\beta = 4 + s, \tag{12}$$

$$L_0 = \frac{16\pi R_0 \sigma T_0^{4+s}}{3\kappa_0 \rho_0^{n+1}}, \tag{13}$$

and all the symbols have their usual meanings.

The three equations (6), (7), and (8) are the equations for the displacement, velocity and temperature respectively and they represent the non-linear radial oscillations of our model.

3. Stability of the Stellar Model

A system of equations (6), (7) and (8) will lead to a characteristic equation, a cubic algebraic equation. It provides information on the solutions of the model. Normalizing the time, i.e., making Ω unity, we have the cubic equation

$$\lambda^3 + A\lambda^2 + B\lambda + C = 0, \tag{14}$$

where

$$A = \epsilon\beta, B = 3\gamma - 1, C = \epsilon(\alpha - \beta), \tag{15}$$

and γ is the ratio of the specific heats and is assumed to be 5/3. There are three roots of which two are not restricted to real numbers. Since we are interested in the oscillatory solutions, we consider the case where (14) has one negative real root and a pair of complex conjugate roots with the positive real part. The condition to avoid the positive real root is $C > 0$ or $\alpha > \beta$, since ϵ is larger than zero. The necessary condition of the pulsational instability is,

$$AB - C < 0. \tag{16}$$

The latter condition indicates the ignition of the well-known κ mechanism. These conditions are used later.

4. Non-Linear Oscillations with the Saturation of the Kappa Mechanism

We study the non-linear oscillations of the model by using numerical integrations of the equations. The normalization of the time is used throughout the calculations. For the parameters which yield the pulsational instability, the orbits of solutions diverge from the origin. With the increase of

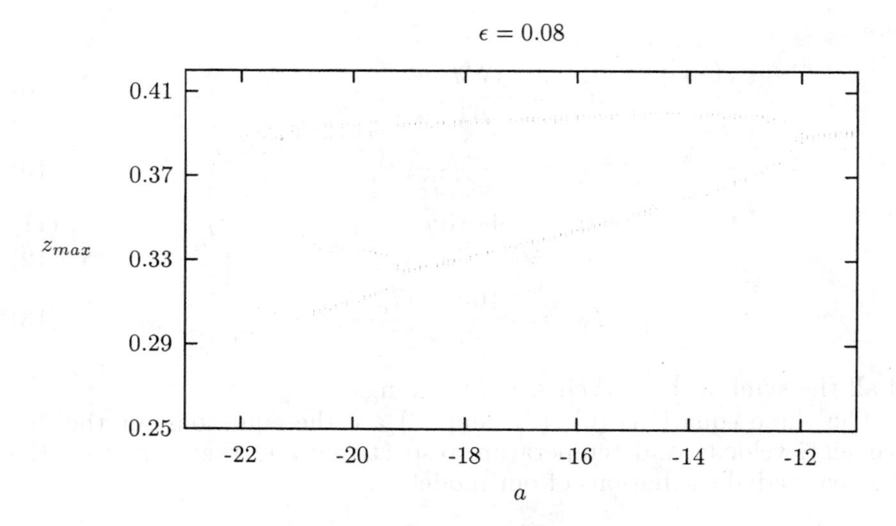

Figure 1. A Feigenbaum tree. The abscissa shows the value of a and the ordinate indicates successive 130 z_{max}. With the decrease of the control parameter a the orbit bifurcates through the period doubling sequence.

the amplitudes of oscillation, asymmetry in the x-z plane becomes clear and the growth of oscillations continues monotonically, so that there is no evidence of any limit-cycle orbit.

To avoid such a simple divergency of the model, Buchler and Regev (1982) assumed non-linearity in the specific heat. In the present, we put the saturation effect into the κ mechanism, by choosing α and β as follows (Devasagayam et al. 1999):

$$\alpha = a\left[(1+x)^{-3}(1+z) - 1.44\right] + 32.14, \tag{17}$$

$$\beta = 3.6(1+x)^{-3}(1+z)\left[(1+x)^{-3}(1+z) - 0.2\right], \tag{18}$$

where a is the parameter to control the κ mechanism. We can see that the temperature derivatives of opacities, $(4-\beta)$, are assumed to be unchanged with the control parameter, while the radius derivatives, $(4-\alpha)$, are assumed to be variable with the parameter a. P is normalized by P_0. As a result of the linearized analysis mentioned above, we may expect the changes of pulsational instability for control parameter a. Although the assumption seems a little unrealistic compared with the real structure of variable star envelopes, we use these relations in the following, since to study the observed property with the changes of the pulsational stability, the instability is essential at the present stage of the investigation.

For the present, we presume $\gamma = 5/3$ and $\epsilon = 0.08$ throughout the following numerical integrations. We show the Feigenbaum tree of the oscillator in Figure 1, where the period-doubling bifurcation with the decrease of the control parameter a is indicated. When the control parameter a is sufficiently large the model is pulsationally stable. With a decrease in the control parameter a, the model becomes pulsationally stable with a period 1 orbit for $a = -11$ in Figure 2, periods 2 for $a = -16$ in Figure 3, periods 4 for $a = -20$ in Figure 4 and chaos appears $a = -22$ in Figure 5.

The return of maps for successive x_{max} show one period for $a = -11$, two periods for $a = -16$, four periods for $a = -20$ and chaotic behavior for $a = -22$. We see the same phenomena is repeated for y_{max} values. If we made a plot for values of $a < -23$, the dots would be spread all over the area in the diagram. Different from the Rossler (1976) oscillator, however, the curve is folded with twist when the control parameter is sufficiently small. With the decrease of the control parameter, the folding process continues. Eventually, the maps show complex patterns. Features of the return maps are a little different from each other for x, y and z, but the pattern of folding is essentially the same.

The return maps are similar to the pictures generated from actual data of five RV Tauri variable stars and four semiregular variable stars by Saitou (1989). The pictures are generated with luminosity maxima on the abscissa and the succeeding maxima on the ordinate of the variable stars.

5. Results and Discussion

We must discuss how the light curves of the one-zone models are to be expressed. Since the surface temperature is regarded as a very small quantity in the Baker (1966) approximation, the luminosity of models is defined as the same quantity as the total radiative flux in the envelope.

As the control parameter values decrease to $a = -22$ there is a clear sign of instability in the curve. Since equation of light is associated with radius, the fluctuation in x influences the light variations. The definition of luminosity used above does not seem adequate to express real envelopes. The emergent flux must be sensitive to the temperature of a layer just inside of the photosphere, but not to that of the deep layers used in the models. The small light maxima appeared in the preceding calculations will be absorbed by the critical layers characterised by a great value of the specific heat. Only the strong outflow at radiation near the minimum radius will be realized in real variable stars. A comparison using the maxima of x or y may be justified (Saitou 1989).

$a = -11; \epsilon = 0.08$

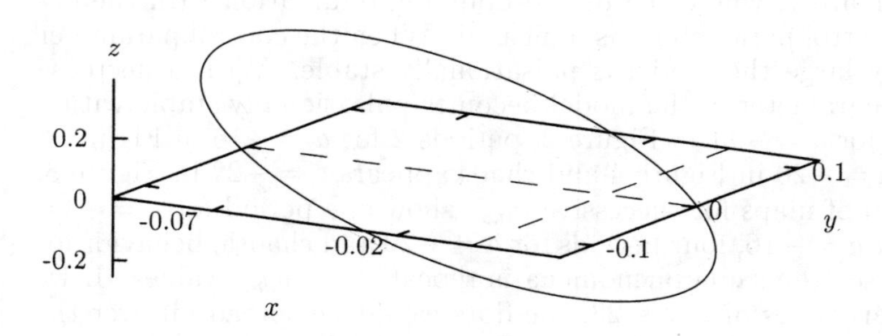

Figure 2. Orbit of the oscillator $\gamma = (5/3)$ and $\epsilon = 0.08$ are chosen for the oscillator demonstrated in the figure $a = -11$.

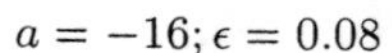

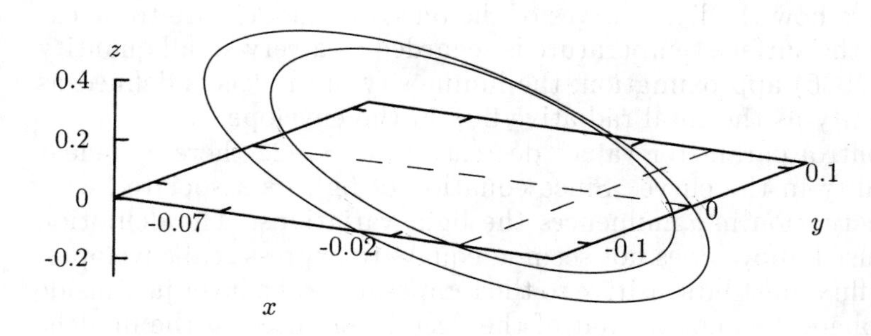

Figure 3. Orbit of the oscillator $\gamma = (5/3)$ and $\epsilon = 0.08$ are chosen for the oscillator demonstrated in the figure $a = -16$.

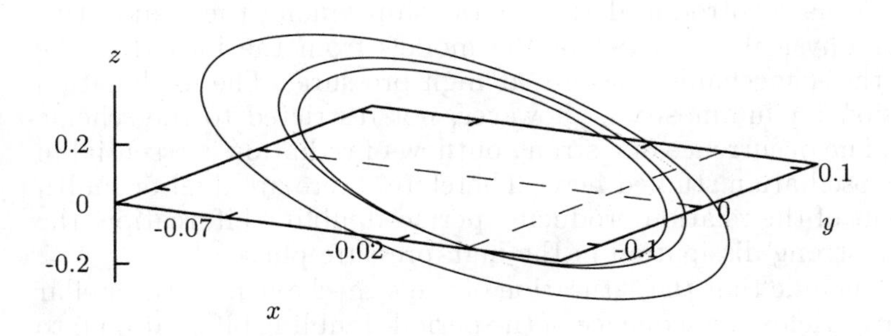

Figure 4. Orbit of the oscillator $\gamma = (5/3)$ and $\epsilon = 0.08$ are chosen for the oscillator demonstrated in the figure $a = -20$.

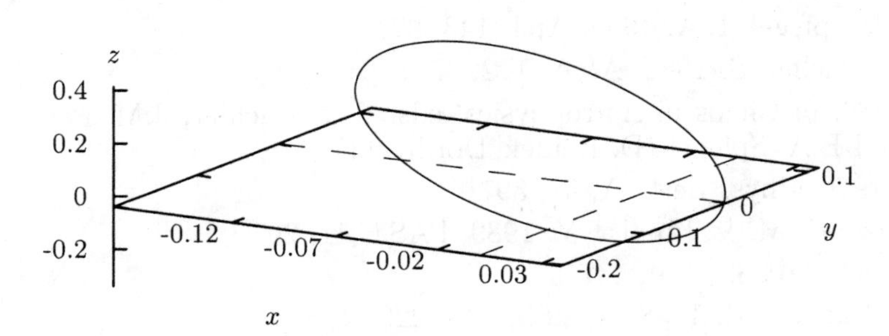

Figure 5. Orbit of the oscillator $\gamma = (5/3)$ and $\epsilon = 0.08$ are chosen for the oscillator demonstrated in the figure $a = -22$.

Confined to the non-linear dynamics of an oscillating system, we do not treat physical or hydrodynamical processes working as the agent for dissipation. We have introduced the relationship among the radioactive flux and other physical quantities of the models from the idea that the saturation of the κ mechanism occurs at high pressure. The explanation of the expression for luminosity is, however, not restricted to the scheme stated above. The occurrence of a strong outflow of radiation is essential in the non-linear oscillation studied here. Therefore, there are dual or multiple explanations of the relation producing period-doubling bifurcation, the saturation of a strong dissipation in the high-pressure phase.

We may conclude that the saturation of the κ mechanism in the stellar envelopes clearly yields the sequence of the period-doubling bifurcation into chaos, although the possibility that other processes in stellar envelopes work as an agent to produce the period doubling cannot be ruled out.

References

Auvergne, M. & Baglin, A. 1985, A&A, 142, 388

Baker, N. 1966, in Stellar Evolution, eds. R.F. Stein and A.G.W. Cameron (Plenum Press: New York)

Buchler, J.R. & Regev, O. 1982, ApJ, 263, 312

Devasagayam, M., Singh, H.P. & Gurm, H.S. 1999, in International Symposium on Astrophysics Research and Science Education, ed. C.D. Impey (Vatican Observatory: Vatican City)

Moore, D.W. & Spiegel, E.A. 1966, ApJ, 143, 871

Perdang, J. & Blacher, S. 1982, A&A, 112, 35

Perdang, J. 1985, in Chaos in Astrophysics, eds. J.R. Buchler, J.M. Perdang and E.A. Spiegel (D. Reidel: Dordrecht)

Rossler, O.E. 1976, Phys. Lett. A, 57, 397

Saitou, M., Takeuti, M. & Tanaka, Y. 1989, PASJ, 41, 297

Takeuti, M. 1988, Ap&SS, 140, 431

Tanaka, Y. & Takeuti, M. 1988, Ap&SS, 148, 229

International Symposium on Astrophysics Research and on the Dialogue between Science and Religion, Vatican Observatory, 2003
C. D. Impey & C. E. Petry, editors

Quantitative Stellar Spectral Classification: The Determination of Fundamental Stellar Parameters

J. García, M.J. Stock, N. Sánchez

Laboratorio de Investigaciones Astronómicas, Universidad del Zulia, Maracaibo, Venezuela.

J. Stock

Centro de Investigaciones de Astronomía, Mérida, Venezuela.

Abstract. We are developing a project to study the vertical structure of the Milky Way, in which some previous results have already been obtained. In this paper, we present a new method for the quantitative spectral classification that it will be used for this purpose. Our results show that it is possible determine absolute magnitudes and intrinsic colors with adequate accurary, based on the equivalent widths of various absorption lines from objective prism spectra. The calibration of the method takes into account B–K type stars, using high resolution slit spectra.

1. Introduction

For the determination of the spatial distribution of different stellar populations at least two steps are required: (1) the complete and unbiased identification of members of the respective population in the field of interest, and (2) a fairly accurate distance determination for each of the members. A combination of spectral classification with photometry can provide the required data, whenever the spectral classification yields intrinsic colors and absolute magnitudes of adequate accuracy. The MK classification has been applied in this sense many times, but it is well known that the precision of the predicted absolute magnitudes falls short of the desired goal. Also, so far the application of the MK system has been restricted to the visual inspection of photographic spectra. Objective prism spectra would be ideal for this type of galactic study because they reach fainter magnitudes and provide data for many stars. Their low resolution only permits the use of strong spectral features, contrary to what the MK classification requires. Also it is now possible to combine objective prism techniques with the use of CCD detectors. Thus the purpose of our work is to test up to which point objective prism and CCD techniques can be used for stellar classification. It is not our purpose to reproduce or improve on the MK system. Instead we shall try to recover stellar parameters directly from measured spectral features.

2. Observations

Our starting point is the *Hipparcos* Catalogue, which provides accurate parallaxes for a large variety of stars. To this we add the library of CCD spectra of nearly 700 stars in the regions 3820–4500 Å and 4780–5450 Å made available to the public by Jones (1999). The Jones library also provides B and V magnitudes. Practically all of his stars are contained in the *Hipparcos* Catalogue. The resolution of the Jones spectra is 0.6 Å, while typical objective prism spectra may have a resolution of about 5.0 Å. The strongest lines were selected, for which an equivalent or pseudo-equivalent width was determined, depending on whether a true continuum could be found or not. Stars with a *Hipparcos* parallax error greater than 20% were eliminated. Thus practically no supergigants are left in the sample. Inspection of the relationship between color and equivalent width made it clear that early and late type stars had to be treated separately, and that it would be necessary to eliminate very early and very late type stars, which are hardly present in the sample in any case. The formation of these different groups was performed on the basis of measured line criteria only (see Stock & Stock 1999). Due to this restriction, the number or remaining O and B spectra was too small to be included. In view of this fact, we started an observing program that concentrated on O and B-type stars. The spectral range covered was from 3950 Å to 5750 Å. The observing list was made up with the *Bright Star Catalogue* in combination with the *Hipparcos Parallax Catalogue*, selecting all O and B-stars brighter than the 6th apparent V-magnitude. A total of 116 stars were observed (Stock et al. 2002). A typical example of a B-spectrum is shown in Figure 1. In all cases, the identification of the selected lines was taken from the Multiplet Table by Moore (1972).

3. Results

Our main purpose is to recover the physical parameters, i.e., the $(B-V)$ colors and the absolute magnitudes, directly from the measured line strengths. For the Jones Library spectra (A–K type), as algorithm we have chosen a second order polynomail with three line widths as independent variables. From the selected lines, 969 groups of three lines can be formed. We tried the above solutions, separately for early and late type stars, with all possible combinations. The results for the optimum combinations are given in Table 1 and 2. These tables show the best combinations for the absolute magnitudes and intrinsic colors.

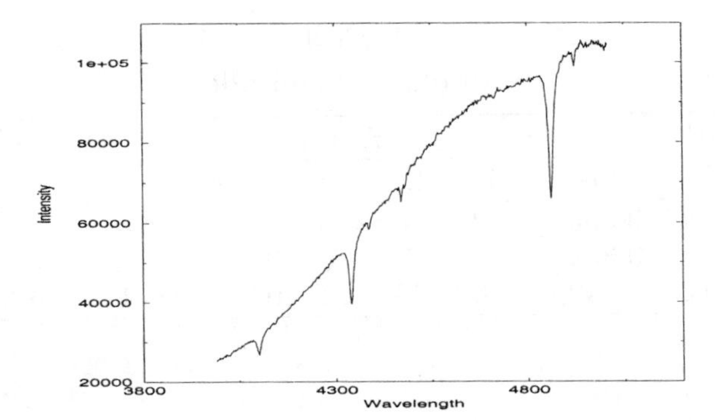

Figure 1. B-type observed spectrum of HD14951. The spectra shows the Balmer lines Hδ, Hγ, and Hβ

In the case of the O and B-spectra, intrinsic colors $(B - V)$ were deduced from the MK-types. The respective relations may be found in tables of Allen (1973). These colors can be compared with the observed colors and a reddening effect can be found. If interpreted as due to absorption by interestellar or circumstellar material, the effect on the apparent V-magnitude can be estimated. For the relation between the physical parameters M_V or $(B-V)$ with the equivalent widths we adopt second order polynomials with two or three independent variables, the latter being the equivalent widths of two or three absorption lines. Thus, the equation for the absolute magnitude with three lines has the form

$$\begin{aligned} M_V = \; & a_{000} + a_{100}w_1 + a_{010}w_2 + a_{001}w_3 + \\ & a_{200}{w_1}^2 + a_{110}w_1w_2 + a_{101}w_1w_3 + \\ & a_{020}{w_2}^2 + a_{011}w_2w_3 + a_{002}{w_3}^2 \end{aligned} \tag{1}$$

where w_1,w_2, and w_3 are the equivalent widths of the three respective absorption lines. The coefficients a_{ijk} have to be determined by least squares, forming equation (1) for all spectra in which values for the three different equivalent widths were obtained. The best combinations for the absolute magnitudes and intrinsic colors are given in Tables 2 and 3, in three groups (O–B, A–F and G–K). $L1$, $L2$, and $L3$ are the equivalent widths of the lines in each combination (for the identification, see Stock & Stock 1999, Stock et al. 2002), while rms is the average residual for each of the combinations.

4. Conclusions

For the Jones library spectra (A–K type stars), algorithms were found which yielded the absolute magnitudes for all spectral types with an average error of 0.26 magnitudes. The $(B - V)$ colors could be reproduced

Table 1
Absolute Magnitude

O–B				A–F				G–K			
L1	L2	L3	rms	L1	L2	L3	rms	L1	L2	L3	rms
1	9	12	0.395	2	4	16	0.263	1	11	16	0.237
3	9	12	0.334	2	7	16	0.265	2	4	5	0.237
9	10	12	0.342	6	11	16	0.267	11	16	19	0.238

Table 2
$(B - V)$ Color

O–B				A–F				G–K			
L1	L2	L3	rms	L1	L2	L3	rms	L1	L2	L3	rms
1	8	11	0.012	2	5	7	0.016	3	5	8	0.019
4	10	12	0.014	3	9	10	0.019	5	6	9	0.020
8	10	12	0.013	5	7	17	0.015	9	17	19	0.016

with an average error of 0.020 magnitudes. In the case of the O–B type stars, the best solutions were obtained with polynomials of three lines, reproducing the absolute magnitudes with an average residual of about 0.40 magnitudes and the intrinsic colors with an average residual of 0.016 magnitudes. The accuracy reached by the calibration in both cases proves the consistency of the method over the entire range of spectral types included in the analysis. This result shows the efficiency of the method as a powerful tool for the study of stellar populations in the Milky Way.

References

Allen, C.W. 1973, Astrophysical Quantities (The Athlone Press: London)

Jones, L.A. 1999, Ph.D. thesis, University of North Carolina

Moore, C. 1972, in Table of Multiplets, National Standards Reference Data Series (National Bureau of Standards: Washington)

Stock, J., & Stock, M.J. 1999, Rev. Mex. A&A, 35, 143

Stock, M.J., Stock, J., García, J. & Sánchez, N. 2002, Rev. Mex. A&A, 38, 129

International Symposium on Astrophysics Research and on the Dialogue between Science and Religion, Vatican Observatory, 2003
C. D. Impey & C. E. Petry, editors

Separating Temperature and Gravity Effects in Brown Dwarf Spectra

Nadya Gorlova

Department of Astronomy, The University of Arizona, Tucson, AZ, USA 85721

Abstract. We analyze J and K-band spectra of a sample of very low mass objects with a wide range of ages in order to investigate the gravity and temperature sensitivity of a few prominent spectral features.

1. Introduction

The existing spectral type classification of the objects near the sub-stellar mass boundary (hereafter Brown Dwarfs or BDs), is primarily based on the behavior of the strong molecular features, like H_2O, TiO, VO, and FeH, in the optical spectra of the nearby field dwarfs. The smooth behavior of these features is the result of the decrease in temperature toward later spectral types, which at some stage includes the process of dust formation.

Recently, many studies of the initial mass function in young clusters revealed a population of very low-mass cool objects, which are observable due to the fact that they are bright at young ages. Due to high extinction in the star-forming regions, one would like to have some spectral criteria in the IR to derive the properties of these objects. One task would be to correlate optical and IR criteria for spectral classification. Another problem arises from the fact the surface gravities of young BDs should be two orders of magnitude lower than those of old, field BDs. Therefore, when one tries to assign a spectral type to the cluster BD by comparing its spectrum with the field standard, the mismatch can be erroneously attributed to *temperature* rather than to a *gravity* difference. Finally, having gravity sensitive indices would allow the separation of background red giants (with even lower gravities) and foreground dwarfs from the low mass cluster members.

2. Observations

In order to address questions outlined in the Introduction, namely, to disentangle effects of temperature and gravity on the IR spectra of BDs, we

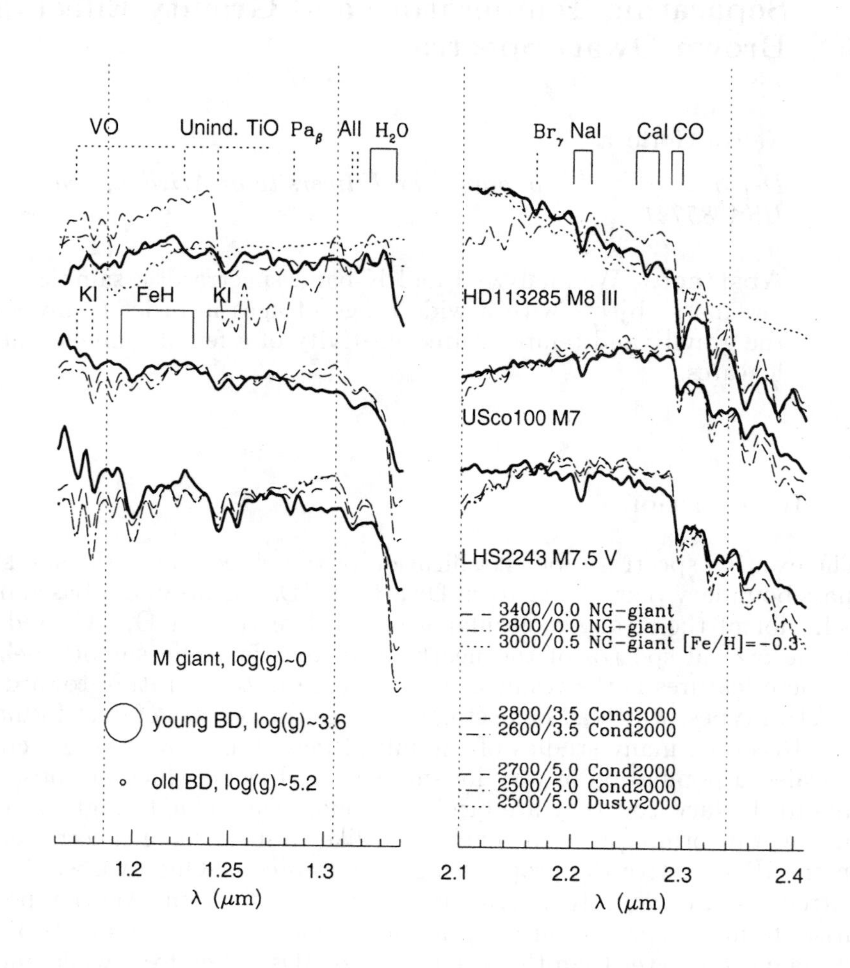

Figure 1. J and K spectra of a giant and two BDs of similar spectral types but different ages, overlaid on the theoretical spectra corresponding to the expected temperature range. Solid lines define indices analyzed in this work for T_{eff} and gravity sensitivity, and dotted lines show other prominent features including those relevant to giants only. As can be seen, KI, FeH and NaI lines are weaker in the younger, bigger BD (USco 100), and so is predicted by the models with lower gravity. H_2O is too strong in the models. Dotted vertical lines show wavelength intervals used for normalization of the spectra to the mean flux.

obtained J and K low-resolution (R $\approx$ 300) spectra of ~30 BDs, spectral types M4–L5, drawn from the field and a few open clusters and associations (ρ Oph, Tau, IC 348, Upper Sco, Pleiades), as well as three late M giants. Most of the objects have known distances, photometry, and spectral types, which allowed us to determine their luminosities and temperatures (T_{eff}), then masses from the theoretical tracks, and finally gravities ($\log g$).

3. Analysis and Results

We detected a few of the most prominent spectral features (Figure 1) and investigated how their strength correlated with T_{eff} and $\log g$.

We confirmed strong temperature dependence of the H_2O and CO bands (eg. Reid et al. 2001) and infer that the gravity dependence of these features is insignificant compared to that of FeH, KI and NaI lines (see also Martin et al. 1996, Luhman et al. 1998, Bejar et al. 1999, and Lucas et al. 2001). Our empirical calibrations are given below (the working range for M6 to M8 was −4 to −2, and for $\log g$, 3.0−5.5 dex).

$$\log g = 2.12(\pm 0.27) + 0.40(\pm 0.05) \times (EW_{KI} \text{Å}) \quad (1)$$

$$\text{Spectral Type} = 13.39(\pm 4.60) - 19.88(\pm 5.70) \times (\text{F}_{1.34}/\text{F}_{1.32}) \quad (2)$$

Where EW_{KI} is the equivalent width of KI line at 1.24 μm and $F_{1.34}/F_{1.32}$ represents our water index, or the ratio of fluxes at 1.34 and 1.32 μm. The precision of the fits can be further improved by increasing the number of objects and the spectral quality.

We also compared our spectra to recent atmospheric calculations by Allard et al. (2001) and concluded that while they predict observable trends correctly, the systematic offsets persist, as best seen at the example of deep water bands in Figure 1. Therefore, current models should be used with caution for the derivation of physical parameters from spectra. Better treatment of dust and fuller opacity tables promise to improve this situation in the near future.

The spectra of late M-giants are very distinctive from the sub-stellar-mass objects: strong NaI, CaI and molecular bands (TiO, VO and CO), except for H_2O. This can be explained by their extremely low gravities, hotter temperature scale and non-equilibrium processes in the tenuous atmospheres.

Acknowledgments. I would like to thank my advisors M. Meyer, G. Rieke and J. Liebert who suggested this project and assisted in many ways in its realization, and the Jesuits of the Vatican Observatory who provided me with a graduate scholarship at the University of Arizona.

References

Allard, F., Hauschildt, P.H., Alexander, D.R., Tamanai, A. & Schweitzer, A. 2001, ApJ, 556, 357

Bejar, V.J.S., Zapatero Osorio, M.R. & Rebolo, R. 1999, ApJ, 521, 671

Lucas, P.W., Roche, P.F., Allard, F. & Hauschildt P.H. 2001, MNRAS, 326, 695

Luhman, K.L., Rieke, G.H., Lada, C.J. & Lada, E.A. 1998, ApJ, 508, 347

Martin, E.L., Rebolo, R. & Zapatero Osorio, M.R. 1996, ApJ, 469, 706

Reid, I.N., Burgasser, A.J., Cruz, K.L., Kirkpatrick, J.D. & Gizis J.E. 2001, AJ, 121, 1710

International Symposium on Astrophysics Research and on the Dialogue between Science and Religion, Vatican Observatory, 2003
C. D. Impey & C. E. Petry, editors

The Association of High-Mass Microblazars with Variable Gamma-Ray Sources

Marina Kaufman Bernadó, Gustavo E. Romero

Instituto Argentino de Radioastronomía, C.C.5, (1894) Villa Elisa, Buenos Aires, Argentina

I. Felix Mirabel

CEA/DMS/DAPNIA/Service d'Astrophysique, Centre d'Etudes de Saclay, 91191 Gif-sur-Yvette, France

Abstract. A model to explain variable gamma-ray sources with high-mass microblazars is described. Inverse Compton interactions between the jet of a microquasar and the UV-photon field of the stellar companion can produce high-energy gamma-rays. The interaction with photons from the disk and the corona is also taken into account. The long-term variability presented by some sources is related to the precession of the jet, which is induced by the gravitational effect of the companion upon the accretion disk. Observations of repeated soft gamma-ray outbursts from Cyg X-1 can be consistently interpreted under this model.

1. Scenario

The third EGRET catalog (Hartmann et al. 1999) contained, when published, ~170 gamma-ray sources that could not be clearly identified with potential counterparts at lower energies. These unidentified sources can be separated into two broad groups. The first one, located at low Galactic latitudes ($|b| < 10°$), is well-correlated with the spiral arms and the Galactic plane. It is formed by bright population I sources (Romero 2001). The other group, at middle or high latitudes, is in turn composed of three different populations: an isotropic component, a group of weak and hard sources probably related to the Gould Belt, and a halo component of soft sources (Grenier 2001). The identification of these sources is one of the most important challenges of contemporary high energy astrophysics.

In this paper we outline a model that can explain many of the characteristics of some low-latitude gamma-ray sources on the basis of presumed *microblazar* activity in distant Galactic microquasars. A microblazar phase occurs when the jet of a microquasar (basically an X-ray binary with relativistic plasma ejections) points nearly toward the observer.

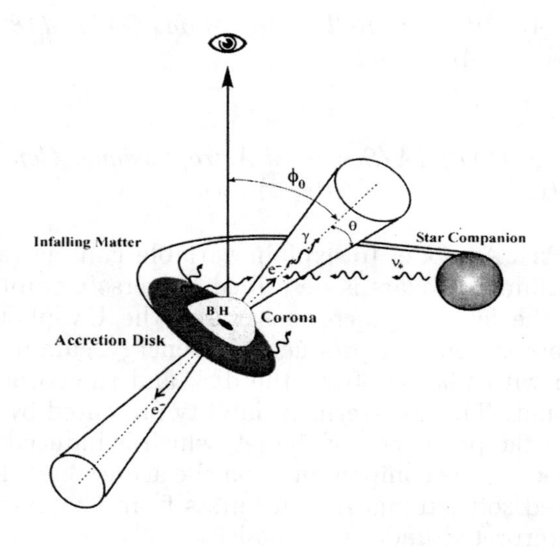

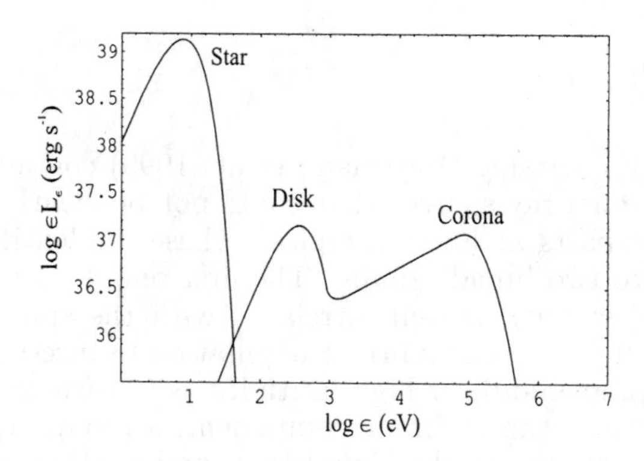

Figure 1. *Upper*: Sketch of the proposed situation. *Lower*: Prototype example of the spectrum from the external photon fields to which the jet is exposed.

We shall argue in this paper that the gravitational effects of the high-mass companion star can induce a precession of the relativistic beam in most of these systems. When the original viewing angle is not too large, the object can go through phases of extremely enhanced activity (the microblazar phase) resulting in a strong gamma-ray source in the observer's frame. The gamma-ray emission is produced by inverse Compton (IC) interactions between relativistic leptons in the jet and seed photons from the star, the accretion disk, and the hot corona or ADAF (advective accretion flow) region around the compact object.

2. Model

As mentioned previously, microquasars are X-ray binary systems able to generate relativistic jets. These jets can be detected through their synchrotron emission at radio wavelengths. We will refer to microquasars as Radio Emitting X-ray Binaries (REXBs). Microblazars are microquasars with jets forming a small angle with the line of sight. They should present a significant enhanced non-thermal flux due to Doppler boosting (Mirabel & Rodríguez 1999).

In what follows we shall consider high-mass microquasars. The jet, which is created close to the central accreting object, must traverse several external photon fields. Figure 1, left panel, sketches this situation, whereas the right panel shows an example of the typical spectra of these fields. First, we have a power-law X-ray isotropic field that can be modeled as a corona or an ADAF region, with a hard spectrum. The IC interactions are in this region under the Klein-Nishina regime for MeV-GeV electrons. Then, we have the field of photons coming from the cold accretion disk, represented by black body emission with a peak at a few keVs. Interactions with this field occur under the Thomson regime. Finally, we have a nearly isotropic field due to the stellar companion, typically a black body with a peak in the ultraviolet. IC interactions occur here also with a Thomson cross section.

For the stellar photons, the resulting IC luminosity, is given by (Georganopoulos et al. 2002):

$$\frac{dL}{d\epsilon d\Omega} \approx D^{2+p} \frac{kV\sigma_T c U 2^{p-1}}{\pi\epsilon_0 (1+p)(3+p)} \left(\frac{\epsilon}{\epsilon_0}\right)^{-(p-1)/2}, \tag{1}$$

where $D = [\Gamma(1-\beta\cos\phi)]^{-1}$ is the Doppler factor, ϕ, the viewing angle, ϵ, the final photon energy, σ_T, the Thomson cross section, U, the energy density of the photon field, k, the constant in the particle energy distribution given by $N(E) \propto E^{-p}$, and p, the spectral index.

Since the disk photons come from behind the jet, we have to add an extra factor $(1-\cos\phi)^{-(p+1)/2}$ that reduces the effects of beaming (Dermer et al. 1992). The IC scattering of the coronal photons is under the Klein-

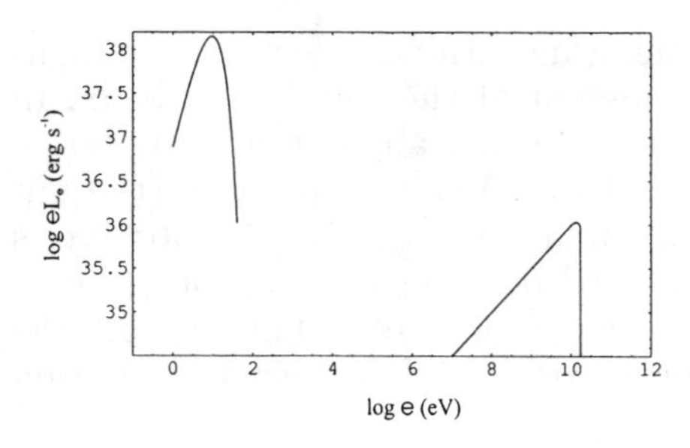

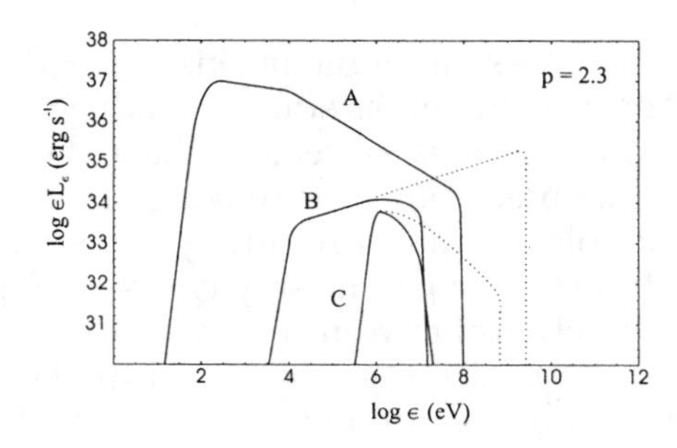

Figure 2. *Left*: Spectral high energy distribution of the scattered photons coming from the companion star. The black body spectral energy distribution of the star is also shown (left top corner) *Right*: Spectral energy distribution for the up-scattering of the star (A), disk (B), and coronal (C) photons. Radiation absorbed in the local photon field is shown by dashed lines.

Nishina regime and then a numerical integration is needed. The result will not be a strict power-law in such a case.

In Figure 2, left panel, we concentrate on the luminosities obtained from the IC of the seed photons from the companion star. It shows the results obtained from the calculation of the spectral energy distribution for a specific model with $p = 2$, bulk Lorentz factor $\Gamma = 5$, viewing angle $\phi = 10°$, photon field of an O7 star (photon energy peaking at ≈ 10 eV), and a high energy cut-off of $\gamma_{max} = 10^{3.5} >> \gamma_{min}$.

The right panel of the same figure shows the result of the up-scattering from the three different seed components: the star, the disk, and the corona or ADAF. In this case, the model calculations are for a specific source, Cyg X-1, taking $p = 2.3$, $\phi = 30°$, $\Gamma = 5$ and $\gamma_{max} = 10^3$ (see Romero et al. 2002 for a detailed treatment of this source). Radiation absorbed in the local photon fields by pair production is shown in dashed lines. Compton losses in the different regions will modify the injected electron spectrum, which will introduce breaks in each spectral energy distribution.

2.1. Variability

The companion star in a high-mass microquasar not only provides a photon field for inverse Compton interactions, but also a gravitational field that can exert a torque onto the accretion disk around the compact object. The effect of this torque, in a non-coplanar system, is to induce a Newtonian precession of the disk. If the jets are coupled to the disk, as is usually thought, then the precession will be transmitted to them. A time-parameterization of the jet's viewing angle is the needed, $\phi(t)$ (Abraham & Romero 1999), which implies a time evolution of the boosting amplification factor of the gamma-ray emission: $D^{2+p} = [\Gamma(1 - \beta \cos\phi(t))]^{-(2+p)}$. The viewing angle, ϕ, and the opening angle of the cone defined by the

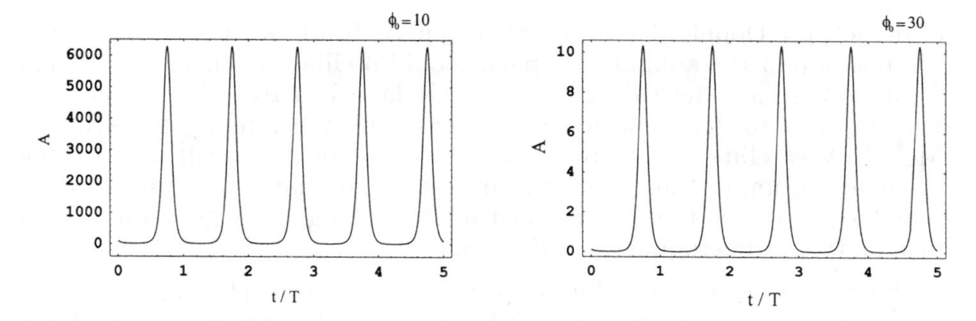

Figure 3. Variation of the amplification factor for a continuous jet emission as a function of time in the precessing microblazar model for two different average viewing angles, ϕ_0= 10 and 30 degrees. Time units are normalized to the precessing period.

jet while precessing, θ, can be seen in the left panel of Figure 1. See Kaufman-Bernadó et al. (2002) for additional details.

The time dependence of the amplification factor is shown in two different examples in Figure 3. The peak of the amplification, when an otherwise undetectable source can fall within the sensitivity of gamma-ray instruments, corresponds to the microblazar phase. It can be seen that a huge range of amplifications, going from a factor 10 to a factor 6×10^3, can be obtained depending mainly on the average viewing angle in the precession, which is of 10 or 30 degrees in the cases depicted in Figure 3.

3. Final Comments

It is clear from Figure 2 that in the case of a high-mass companion microblazar, the IC photons from the star dominates the observed emission. As we can see, energies in the range MeV-GeV can be reached for reasonable electron Lorentz factors of $\gamma \sim 10^3$. Nevertheless, energies of a few MeV corresponding to soft-gamma emission or hard X-rays, can also be obtained with this model, as is shown here in the specific case of Cyg X-1 (see also Georganopoulos et al. 2002), or recently suggested by Zhang et al. (2002) for the new unidentified MeV source detected by COMPTEL. The calculated luminosities are easily in the expected range 10^{35}-10^{36} erg/sec required for EGRET sources in the Galactic plane. The resulting photon energy distribution, and the luminosity inferred by the observer, will strongly depend on the assumed spectrum for the relativistic electrons of the jet and on the viewing angle ϕ which will determine the degree of "microblazar" of the studied microquasar.

Most of the gamma-rays produced within the coronal region will be absorbed by pair creation. The annihilation of these pairs would produce a broad, blue-shifted feature in the MeV spectrum. Because of the precession of the jet, the Doppler factor will change periodically with time, and hence the position of the annihilation peak should oscillate in energy in the lab frame around a mean value. The recently launched INTEGRAL satellite may be able to detect such features. On the other hand, the GLAST MeV-GeV satellite, due to be launched in 2005 or 2006, will enhance the statistics on unidentified Galactic sources making possible more accurate population studies that might shed additional light on the number and distribution of microquasars in the Galaxy.

Acknowledgments. High-energy astrophysics with GER is supported by Funadación Antorchas, with additional contributions from CONICET (PIP 0430/98) and ANPCT (PICT No. 03-04881). This work benefited from the Argentinian-French ECOS-SUD cooperation agreement. MK-B and GER thank the kind hospitality of the Service d'Astrophysique, Centre d'Etudes de Saclay. MK-B is grateful to the organizing committee of SuperVOSS II for financial help to allow participation in the meeting.

References

Abraham, Z. & Romero G.E. 1999, A&A 344, 61

Dermer, C.D., Schleickheiser, R. & Mastichiadis, A. 1992, A&A 256, L27

Georganopoulos, M., Aharonian, F.A. & Kirk, J.G. 2002, A&A, 388, L25

Grenier, I.A. 2001, in The Nature of Unidentified Galactic Gamma-Ray Sources, eds. A. Carraminana, O. Reimer and D. Thompson (Kluwer Academic Publishers, Dordrecht), p. 51

Hartmann, R.C., Bertsch, D.L., Bloom, S.D. et al. 1999, ApJS, 123, 79

Kaufman-Bernadó, M.M., Romero, G.E. & Mirabel, I.F. 2002, A&A, 385, L10

Mirabel, I.F. & Rodríguez, L.F. 1999, ARA&A 37, 409

Romero, G.E. 2001, in The Nature of Unidentified Galactic Gamma-Ray Sources, eds. A. Carraminana, O. Reimer and D. Thompson (Kluwer Academic Publishers, Dordrecht), p. 65

Romero, G.E., Kaufman Bernadó, M.M. & Mirabel, I.F. 2002, A&A, 393, L61

Zhang, S., Collmar, W. & Schönfelder 2002, A&A, 396, 923

International Symposium on Astrophysics Research and on the Dialogue between Science and Religion, Vatican Observatory, 2003
C. D. Impey & C. E. Petry, editors

Probing Rare Stages of Stellar Evolution in NGC 5128 Globular Clusters

Dante Minniti

D. Astronomia, P. Univ. Catolica, Casilla 306, Santiago 22, Chile

Marina Rejkuba

ESO, Karl-Schwarzschild-Str. 2, D-85748 Garching, Germany

Abstract. We use VLT images to search for globular clusters in the inner regions of NGC 5128, which are very reddened and crowded. About a hundred new clusters were identified on the basis of their sizes, ellipticities, colors, and magnitudes. We started a spectroscopic survey of the candidates, finding one globular cluster with strong narrow emission lines. These lines are typical of a planetary nebula, and on the basis of the spatial coincidence and of the radial velocity match, we argue that we have found the first planetary nebula that belongs to a globular cluster of NGC 5128. Finally, we discuss the difficulties expected for finding other similar objects.

1. NGC 5128: The Nearest Giant Elliptical Galaxy

The southern E/S0pec galaxy NGC 5128 is the closest giant elliptical galaxy, at about 4 Mpc (Rejkuba 2002). It is about 10 times more massive than the Milky Way, extending over 2 degrees in the sky. NGC 5128 contains 10 times more globular clusters than our own galaxy (Kissler-Patig 1997). The globular clusters are the oldest populations known in galaxies, and have been used as laboratories to test stellar evolution models. These are luminous objects, very conspicuous in the halos of galaxies like NGC 5128. However, globular clusters in the inner regions are more difficult to find, yet the population deep in the potential well of the giant galaxies is very interesting in terms of the dynamics, chemical enrichment, and the formation history of these systems. We discuss in the next section a search for globular clusters in the inner 4 kpc of NGC 5128, a region of patchy dust obscuration and high surface brightness.

In addition, some rapid stages of stellar evolution that are rare in the Milky Way globulars can be better studied in galaxies that contain large numbers of globular clusters, like NGC 5128. Our data allow us to study one particular example of a rare stage of stellar evolution: globular cluster planetary nebulae. Indeed, globular cluster planetary nebulae are

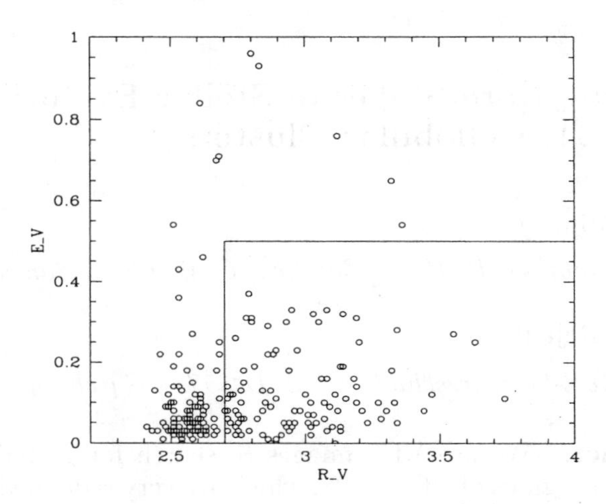

Figure 1. Ellipticities, b/a, versus sizes in pixels for sources in the NGC 5128 inner field. The globular cluster candidates should lie inside the box, and unresolved point sources (stars) appear with a mean size of 2.55 pixels.

very rare, only 4 such objects have been found in the Milky Way globular cluster system (Jacoby et al. 1997). Section 3 discusses the discovery of one such object in NGC 5128, along with a description of the search for globular cluster planetary nebulae in general.

2. Selection of Globular Clusters

BVI images of NGC 5128 taken with FORS2 at the VLT were obtained from the ESO Archive. These VLT images were used to identify candidate globular clusters in the central 4 arcmin region of this galaxy. Based on the sizes, optical magnitudes and colors, a hundred new candidate globular clusters were identified with projected galactocentric distances 1 kpc $<$ $R_G < 5$ kpc. These clusters have magnitudes $18 < V < 20.5$, or $-10.3 <$ $M_V < -7.8$ at the distance of this galaxy, and colors $0.8 < (V - I) < 1.5$, and $0.5 < (B - V) < 1.5$.

NGC 5128 is close enough to allow the identification of globular clusters based on their sizes (Minniti et al. 1996, Rejkuba 2001): at $D = 4$ Mpc the typical clusters should be resolved from the ground under good seeing conditions ($<$ 1 arcsec). In particular, Rejkuba (2001, 2002) used the ESO VLT to find clusters in this galaxy based on their sizes in the U and V-bands. She discovered 80 new globular clusters in two halo fields at 9 and 15 kpc from the center of NGC 5128.

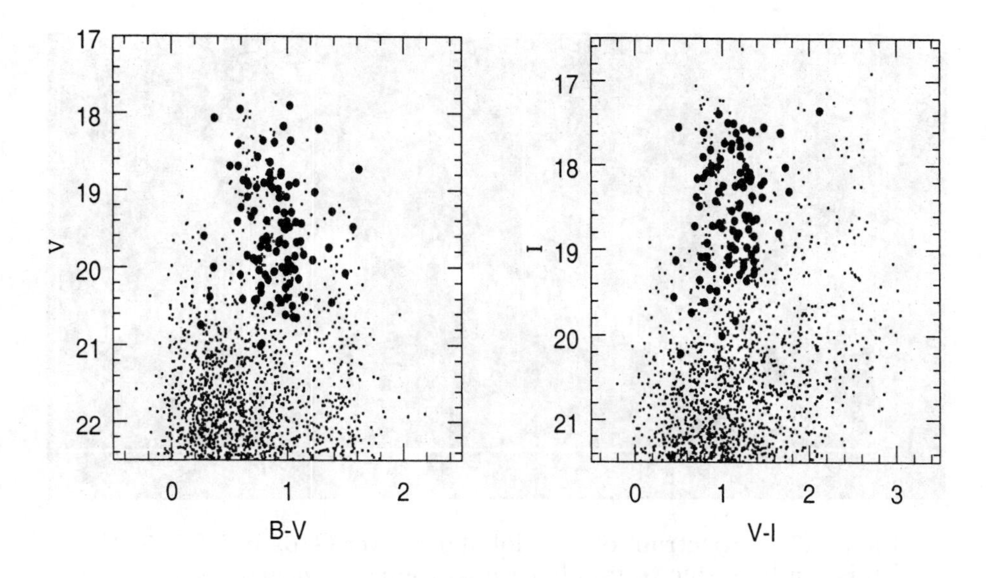

Figure 2. Color-magnitude diagrams of the selected clusters (dark circles), compared with the rest of the sources in the field.

We measured the sizes and shapes of sources in the VLT images, in order to discriminate globular clusters from foreground stars and background galaxies (see Minniti et al. 1996). Figure 1 shows the size-ellipticity diagram. The sizes R_V are the full-widths at half maximum in pixels measured directly from the V-band images. Based on this figure, we select the objects with $R_V > 2.7$ pixels and $E_V < 0.5$ as globular cluster candidates. Instead, the stellar sources have a mean size of $R_V = 2.5$ pixels, or 0.5 arcsec, with a very small spread ($2.4 < R_V < 2.6$ pixels).

The candidate objects need to have colors and magnitudes corresponding to globular clusters. Figure 2 shows the V versus $B - V$ and I versus $V - I$ color-magnitude diagrams for the objects plotted in Figure 1. Based on the sizes, shapes, colors and magnitudes, we secured a sample of a hundred new globular clusters brighter than $M_V = -7.5$ in the inner 4 kpc of NGC 5128. This magnitude corresponds to the peak of the globular cluster luminosity function, our sample is incomplete for fainter candidates, which do not have reliable sizes measured. Finally, we note that some of these new globular clusters have been identified with bright CHANDRA X-ray point sources (Minniti et al. 2002).

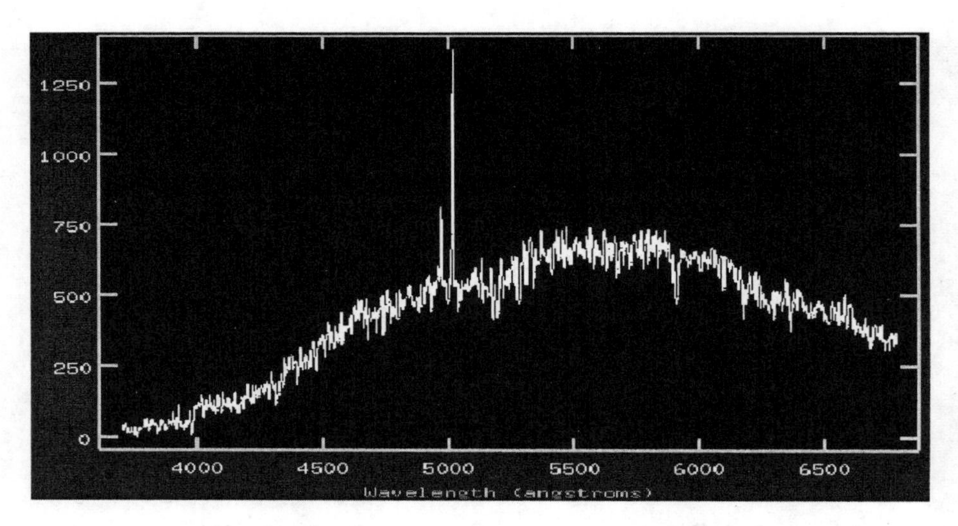

Figure 3. Spectrum of the globular cluster G169 in NGC 5128. Emission lines due to its planetary nebula are clearly seen.

3. Globular Cluster Planetary Nebulae

In order to test the reliability of our clusters, a spectroscopic follow up survey was started with the Magellan I Baade 6.5 m telescope at Las Campanas Observatory using the Boller & Chivens spectrograph. During the first run, a few dozen of the inner NGC 5128 globular clusters selected from the VLT photometry were observed. Much to our surprise, we noticed at the telescope that one candidate cluster exhibited narrow emission lines (Figure 2), in contrast with the other "normal" clusters. This object, located at $\alpha = 13^h25^m30\fs1, \delta = -42°58'11''$ (J2000) turns out to be a well known NGC 5128 globular cluster: it is cluster G169 of Sharples (1988), and also cluster 19 of Minniti et al. (1996).

The globular cluster G169 has $V = 21.15 \pm 0.01$, corresponding to a $M_V = -7$ globular cluster at the distance of 4 Mpc (ignoring the correction due to the emission lines, and adopting the field reddening from Rejkuba 2001). The metallicity of globular cluster G169 was measured by Harris et al. (1992) using the Washington filters: $[Fe/H] = 0$, and by Minniti et al. (1996) using optical HST images and ground-based near-IR images: $[Fe/H] = -0.18$. Thus, G169 is a typical metal-rich globular cluster.

The underlying background due to NGC 5128 itself was subtracted to produce the spectrum of G169 shown in Figure 3. Aside from a couple of narrow emission lines, it is clear that Figure 3 depicts the spectrum of a typical globular cluster with integrated spectral G-type, showing absorption due to the Mg 5170 triplet, the Ca H & K doublet, the NaD doublet, as well as iron lines like Fe 5270 and Fe 5335. The H Balmer lines are weak, presumably being filled in by the nebular emission. The two most prominent emission lines are [OIII] 5007 and [OIII] 4959, although a few weaker emission lines are also present, and we conclude that they are due to the presence of a planetary nebula.

Sharples (1988) measured a velocity for G169 of $V = 697$ km s^{-1}. Using the [OIII] lines we measure a line of sight velocity of $V = 702$ km s^{-1}, in excellent agreement with the previous value. Based on the spatial coincidence and the velocity match, we conclude that the planetary nebula belongs to the cluster G169 (Minniti & Rejkuba 2002). We also note that the emission line intensities are similar to the spectra of the NGC 5128 planetary nebulae observed by Walsh et al. (1999).

3.1. Searching for GCPN in Distant Galaxies

The meager data on globular cluster planetary nebulae available today seem to suggest that these objects are rarer than expected, from both stellar evolution or local samples. However, before jumping to conclusions, larger samples of extragalactic planetary nebulae and globular clusters must be carefully searched and analyzed.

It is surprising that no extragalactic globular cluster planetary nebulae have been discovered before. This underscores the difficulties in finding them. In order to see what to expect, we performed a simple simulation. Suppose we put all 150 MW globular clusters at the distance of NGC 5128, and that each cluster has 2 planetary nebulae. We take the luminosities of these planetary nebulae to be typical luminosities of NGC 5128 planetary nebulae, randomly drawn from the sample of Hui et al. (1993). As the detectability of a globular cluster planetary nebulae would depend on both the planetary nebulae luminosity and the GC luminosity, we plot in Figure 4 a globular cluster planetary nebulae color-magnitude diagram of $M5007$ verus $V - M5007$.

The selection effects in searching for these objects would be important. The parameter space shown in Figure 4 has four regimes: 1. bright cluster, bright planetary nebula (upper left region of Figure 4); 2. bright cluster, faint planetary nebula (lower left region); 3. faint cluster, bright planetary nebula (upper right region); and 4. faint cluster, faint planetary nebula (lower right region). In regime 1, globular cluster planetary nebulae should be detected. Such an object should be prominently present in the globular cluster samples. In fact, G169 belongs to this class, corresponding to a bright planetary nebula in a bright globular cluster. In regime 2, the faint emission lines of the planetary nebulae would be missed in the bright globular cluster spectrum, unless very high S/N or very high dispersion spectra are available. In regime 3, the globular cluster planetary nebulae could be missed in the planetary nebulae samples, depending on the depth of the continuum frames used to select planetaries. The faint objects of regime 4 would be difficult to identify in a distant galaxy.

Of course the case presented in Figure 4 is not realistic: the presence of a planetary nebula in a globular cluster should be proportional to

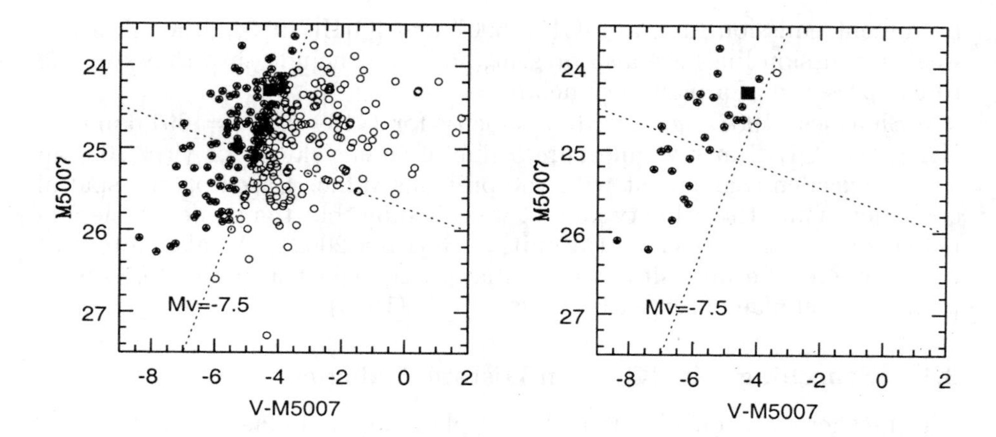

Figure 4. The left panel is a color-magnitude diagram of globular cluster planetary nebulae, simulated by randomly adding two planetaries drawn from the NGC 5128 planetary nebulae population to every single MW globular cluster. There are 300 objects plotted. The dotted lines divide the different regimes discussed in the text. The location of the first object discovered in NGC 5128 is plotted with the dark square. The right panel is a more realistic diagram, simulated by randomly adding planetaries drawn from the NGC 5128 planetary nebulae population following the proportion found in MW globular clusters. The planetary nebulae are now added proportionally to the cluster total luminosities.

the total number of cluster stars, i.e. its luminosity. Thus, the last two regimes discussed above would be very rare, as the number of planetary nebulae is proportional to the globular cluster luminosity. For example, a bright massive cluster like NGC 5139 (ω Cen) with $M_V = -10.3$ is 100 times more likely to have a planetary nebulae than a low mass cluster like NGC 6540 with $M_V = -5.3$. Thus, we repeat the simulation in Figure 6, but now considering a more realistic case. We also use the fact that only 4 Milky Way globular clusters have planetary nebulae (Jacoby et al. 1997), mantaining this frequency for the population of 1700 globular clusters of NGC 5128. Based on Figure 6, we can argue that the objects in the lower left region (regime 2) will not be detected, leaving only 16 globular cluster planetary nebulae to be found in the upper left region (regime 1). These correspond to bright planetary nebulae inhabiting massive globular clusters (or clusters brighter than the peak of the globular cluster LF at $M_V = -7.5$). Unfortunately, even though objects in the upper right region (regime 3) would be easily detected, they are less likely to occur because this region corresponds to low mass clusters.

We note that the most luminous planetary nebulae in globular clusters could be multiple sources instead of single planetaries. Assuming Poisson statistics, the probability of observing a multiple source increases with the number N of sampled objects. In our case, taking the probability of finding a planetary nebulae in a globular cluster as $p = 0.03$ (from the Milky Way), leads to the prediction that fewer than one bright emission line sources in NGC 5128 globulars would be composed of two planetaries.

As planetary nebulae are common in the field halo population, chance superposition of a field planetary with an unrelated globular cluster has not been considered in our simulations, and needs to be carefully evaluated also. This effect would be particularly important in the inner regions of distant galaxies, and it would lead to an overestimate of the population of globular cluster planetary nebulae.

However, the data and simulations suggest that the observed frequency of globular cluster planetary nebulae is less than expected by a factor of 5-10. The scarcity of these objects would suggest that Pop II globular cluster planetary nebulae are rarer than expected from the canonical lifetime $t = 25,000$ yr, but effects such as those discussed above need to be evaluated in detail in order to better constrain the lifetimes of planetary nebulae.

Acknowledgments. We are very grateful to George Coyne and deeply indebted with the Vatican Observatory for the VOSS 86 and VOSS 97, which impacted our lives and launched our careers in Astronomy. DM is supported by FONDAP Center for Astrophysics 15010003.

References

Harris, G.L., Geisler, D., Harris, H.C. & Hesser, J.E. 1992, AJ, 104, 613

Hui, X., Ford, H.C., Ciardullo, R. & Jacoby, G.H. 1993, ApJ, 414, 463

Jacoby, G.H., Morse, J.A., Fullton, L.K., Kwitter, K.B. & Henry, R.B.C. 1997, AJ, 114, 2611

Kissler-Patig, M. 1997, A&A, 319, 83

Minniti, D., Alonso, M.V., Goudfrooij, P., Jablonka, P. & Meylan, G. 1996, ApJ, 467, 221

Minniti, D. & Rejkuba, M. 2002, ApJ, 575, L59

Minniti, D., Rejkuba, M., Funes, J.G. & Akiyama, S. 2002, ApJ, submitted

Rejkuba, M. 2001, A&A, 369, 812

Rejkuba, M. 2002, PhD thesis, Pontificia Universidad Catolica

Sharples, R. 1988, in IAU Symp. 126, Globular Cluster Systems in Galaxies, eds. J.E. Grindlay & A.G.D. Philip (Kluwer: Dordrecht), p. 245

Walsh, J.R., Walton, N.A., Jacoby, G.H. & Peletier, R.F. 1999, A&A, 346, 753

International Symposium on Astrophysics Research and on the Dialogue between Science and Religion, Vatican Observatory, 2003
C. D. Impey & C. E. Petry, editors

Accretion onto Neutron Stars

Sarah Natalia Sägesser

Department of Theoretical Physics, University of Oxford, Keble Road, Oxford OX1 3NP, UK

Abstract. We investigate the possiblity that the asymmetric fall-back on neutron stars of material previously ejected in the pertaining supernova can be responsible for the high pulsar velocities measured. In all models considered the NS will be non-rotating and unmagnetized. The mass infall rates will be high, ($\dot{M} \gg \dot{M}_{Edd}$), the situation is therefore hypercritical. From simulations we conclude that the mass infall rate is the decisive parameter and for $\dot{M} > 10^6 M_\odot$ yr^{-1} neutron stars can indeed receive sufficiently strong kicks to account for spatial velocities of the order of several hundred km s^{-1}.

1. Introduction

At present, despite the difficulties in dealing with selection effects, there can be no question that pulsars have spatial velocities in the range of several hundred km s^{-1} (Hansen 1996). Most authors follow the arguments of Lyne & Lorimer (1994), who interpreted the selection effects in such a way as to obtain an average pulsar velocity of 450 ± 90 km s^{-1}.

There are three classes of explanations for the high pulsar velocities measured, the binary break-up theory, the class of natal kick theories, and the theory that post-natal kicks are responsible is also still under discussion. The binary break-up has recently been discussed by Iben & Tutukov (1961), and although it is a viable alternative for many slower pulsars, it cannot be responsible for velocities of the order of 1500 km s^{-1}.

Asymmetries in type II supernova explosions are the most probable cause of the high space velocities of radio pulsars. Many direct observations of nearby supernovae and supernova remnants support the notion that supernova explosions are *not* spherically symmetric. (e.g. Gaensler et al. 1997) The idea that a non-radial asymmetry could be responsible for the high pulsar velocities is attributed to Shklovskii (1970a). Since 99% of the neutron star energy (a few times 10^{53} ergs) is released in neutrinos, a fractional asymmetry in the radiated neutrino energy of only 1% could in theory generate kick velocities of the order of hundreds of km s^{-1}!

However, the precise origin of these asymmetries is as yet unknown. Recent work has stressed the potential importance of local instabilities occuring subsequent to the collapse of the stellar core. It is also plausible that a substantial asymmetry is already present during core collapse. Asymmetries may stem from convection during the delayed explosion (Herant et al. 1994), from anisotropic neutrino emission (Lai 1998a,b), or even be associated with the g-mode oscillations in the pre-supernova star.

All in all, the single assumption that neutron stars receive an intrinsic natal kick explains a variety of observations (Dewey & Cordes 1987) including spin-orbit misalignment and velocities higher than 1000 km s^{-1} (Burrows & Hayes 1996), and most authors agree that binary break-up cannot be solely responsible for the high pulsar velocities measured. In this work, we assume that an initial asymmetry in the ejecta is already present and we investigate the consequences of the subsequent fallback, particularly on the spatial velocity of the neutron star.

2. Model and Assumptions

During the supernova explosion, by far the largest part of the material is ejected into space at average speeds of around 7000 km s^{-1}. Some of this less violently ejected material, however, does not reach escape velocity and falls back onto the neutron star. Due to small asymmetries in the explosion, the motion of the neutron star and the stellar surroundings, the ejecta is expected to fall back asymmetrically (Tauris & Sennels 2000).

If we assume a total infalling mass $M_{infall} \approx 0.1\ M_{\odot}$, which falls down onto the neutron star in free fall ($v_{ff} \approx 2/3c$) and is distributed according to an asymmetry α in the density of 1%, then with momentum conservation the velocity of the neutron star can be estimated as follows:

$$v_{NS} = \alpha \cdot v_{ff} \cdot \frac{M_{infall}}{M_{NS}} \approx 1400 \text{ km s}^{-1} \quad (1)$$

This simple estimation shows it is possible that the high neutron star velocities measured stem from the fallback of matter.

For hypercritical accretion rates, the trapping radius is several orders of magnitude larger than the neutron star radius. As a result, most of the accretion energy is released inside the trapping radius and cannot be carried away by photons as the matter is optically thick. Instead, the energy is carried away by *neutrinos* produced near the neutron star surface. As material flows toward the neutron star inside the trapping radius, adiabatic compression of the radiation pressure-dominated gas brings the temperature above the pair production threshold. In a thin layer on the neutron star surface, pair annihilation produces neutrinos which immediately escape, carrying away the accretion energy without impeding the flow of material onto the central object.

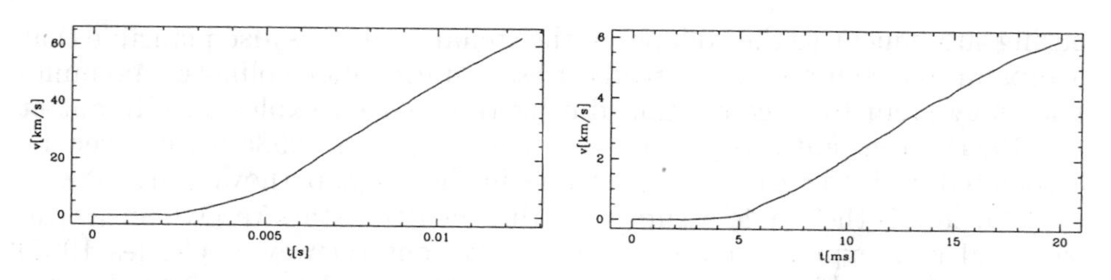

Figure 1. Acceleration for $\dot{M} = 1\ M_\odot\ s^{-1}$ (left) and $\dot{M} = 0.1\ M_\odot\ s^{-1}$ (right).

3. Simulations

We varied the configuration for each simulation, always using the following basic picture: the neutron star ($M_{NS} = 1.4\ M_\odot$, $R_{NS} = 10$ km) is surrounded by an inner hull consisting of a hot, high entropy component, and an outer hull, cooler and denser, and almost in free fall onto the neutron star. We generally started out with 20,000–30,000 particles. To keep the mass accretion rate constant in time, we injected particles at the outer rim (≈ 250 km) at regular intervalls. We typically worked with infall rates of $(10^6 - 10^8)\ M_\odot\ yr^{-1}$. After setting up the particles on their shells, the density perturbation and the infalling mass rate could be determined.

We use a 3D hydrodynamics code based on the SPH method written by W. Benz. The time integration is done with a predictor-corrector scheme. The version of the code used includes gas pressure, dissipation (via shocks) and self-gravity. In our calculations the equation of state is in principle given via a table based on the calculations done by Lattimer & Swesty (1991). Dissipation in shocks is treated using an artificial viscosity. The simulations work with optically thick matter and with a constant electron density $Y_e = 0.5$, and the following neutrino processes have been included: plasma and pair processes (neutrino cooling), absorption by nuclei and scattering by electrons and positrons (neutrino heating).

4. Results

The velocity of the neutron star is calculated from the acceleration due to the recoil by neutrino emission. In all our simulations a generic asymmetry-recoil correlation can be ascertained. However, the magnitude of the kick depends markedly on the chosen input parameters, mainly on the rate of infall, but also on the entropy and density distribution of the neutron star surroundings. In Figure 1 we can see the velocity increase for typical simulation with two different mass infall rates $\dot{M}$. If we assume this process goes on until a total mass infall of $0.1\ M_\odot$ is reached, then after the pertaining time of 100 milliseconds (for $\dot{M} = 1\ M_\odot\ s^{-1}$) and 1 second, respectively, the extrapolated velocity is of the order of several hundred km s^{-1}. For infall rates lower than $10^6\ M_\odot\ \mathrm{yr}^{-1}$, the impact on the neutron star velocity rapidly declines. Changing parameters other than the infall rate does not greatly influence the magnitude of the neutron star velocity.

5. Discussion and Conclusion

Woosley and Weaver (1995) give an initial fallback rate of $10^7\ M_\odot\ \mathrm{yr}^{-1}$ for a 25 $M_\odot$ supernova progenitor. The initial rate of accretion, following the supernova explosion, is estimated by Brown (1995) to be $\dot{M} \cong 1\ M_\odot\ t^{-1}\ \mathrm{s}^{-1}$ with t measured in seconds. Thus, the infall rate 1 second after the supernova explosion is equal to $3 \times 10^7\ M_\odot\ \mathrm{yr}^{-1}$, which would justify the mass infall rates we worked with. However, it seems to us that more precise calculations are asked for, also to determine how exactly initial asymmetries translate into quantities such as fallback rates and the density distribution. We conclude that if infall rates of $10^6\ M_\odot\ \mathrm{yr}^{-1}$ or higher occur after supernova explosions *and* the accretion process is halted before the neutron star collapses to a black hole, then the fallback kick mechanism presented in this thesis is a plausible explanation for even the highest known neutron star velocities.

References

Brown, G.E. 1995, ApJ, 440, 270
Burrows, A. & Hayes, J. 1996 Phys. Rev. Lett., 76, 352
Dewey, R.J. & Cordes, J.M. 1987, ApJ, 321, 780
Gaensler, B.M. et al. 1997, ApJ, 479, 845
Hansen, B.M.S. 1996, PhD thesis
Herant, M. et al. 1994, ApJ, 435, 339
Iben, I. & Tutukov, A.V. 1996, ApJ, 456, 738
Lai, D. 1998a, astro-ph/9712043
Lai, D. 1998b, astro-ph/9802344
Lattimer, J.M. & Swesty, F.D. 1991, Nuclear Physics A, 535, 331
Lyne, A.G. & Lorimer, D.R. 1994, Nature, 369, 127
Shklovskii, I.S. 1970, Soviet Astronomical, 13, 562
Tauris, T.M. & Sennels, T. 2000, A&A, 355, 236
Woosley, S.E. & Weaver, T.A. 1995, ApJS, 101, 181

International Symposium on Astrophysics Research and on the Dialogue between Science and Religion, Vatican Observatory, 2003
C. D. Impey & C. E. Petry, editors

Stroboscopic Optical Observations of the Crab Pulsar

Simon Vidrih, Andrej Čadež, Mirjam Galičič

Department of Physics, Faculty of Mathematics and Physics, University of Ljubljana, Jadranska 19, 1000 Ljubljana, Slovenia

Alberto Carramiñana

I.N.A.O.E., Luis Enrique Erro 1, Tonantzintla, Puebla 72840, México

Abstract. Photometric data of the Crab pulsar, obtained in stroboscopic mode over a period of more than eight years, are presented here. The applied Fourier analysis reveals a faint 60 second modulation of the pulsar's signal, which we interpret as a free precession of the pulsar.

1. Introduction

Among all known pulsars, the Crab pulsar is the brightest in all the spectral regions and, therefore, by far the most thoroughly investigated. It is ~ 1000 years old, with a rotation period of $\sim$ 33ms. In the optical, it is a 16.5 magnitude star and its pulsations were detected more than 30 years ago (Cocke et al. 1969). We observed the Crab pulsar optically in stroboscopic mode for a period of more than eight years. We detected a faint peak in Fourier spectra of the pulsar's light curve at frequency $\sim 1/60$ Hz and suggested that it may be the signature of free precession of the pulsar. Another statistical analysis applied on the extensive data set confirms the presence of a periodic signal with the period of 60 seconds.

2. Stroboscopic Observations of the Crab Pulsar

Our stroboscopic system is based on a shutter that opens with the prescribed frequency and phase. The light signal is thus observed periodically only during one part of the period. We have applied this technique in performing highly accurate phase-resolved optical photometry of the Crab pulsar.

The phase light curve of the Crab pulsar is well known (see Figure 1) and the arrival times of the pulses can be precisely calculated. The

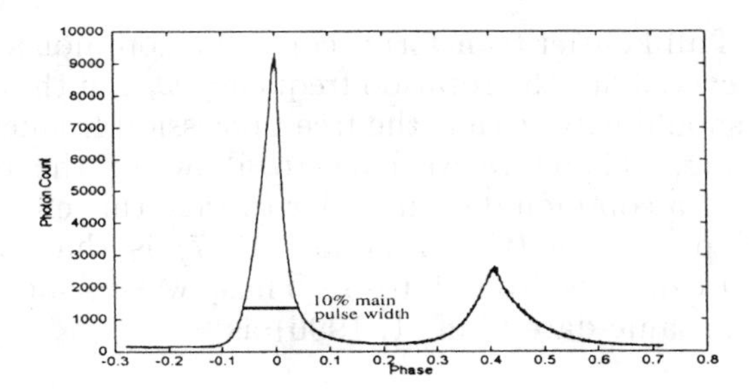

Figure 1. The phase light curve of the Crab pulsar as derived from data taken with the fast speed photometer on board the Hubble Space Telescope (Percival et al. 1993).

shutter is a rotating wheel with the width of the cut-outs corresponding to the width of the Crab pulsar's main pulse. While observing, we lock our stroboscopic shutter in phase to the main peak, so that the CCD detector receives all light from the main pulse, but the bright emission from the surrounding nebula is reduced by a factor of 10, which substantially lowers the noise in the pulsar's signal. In the stroboscopic mode we take several hundred consecutive images with exposure times of several seconds and thus obtain, in each observing run, the pulsar's light curve $m_k(t_l)$, where t_l are consecutive starting times of exposures in the kth observing run.

A faint 60 second periodic modulation of the Crab pulsar light curve was already identified seven years ago (Čadež & Galičič 1996) and a simple free precession model was proposed to explain the observed phenomenon (Čadež et al. 1997). We have continued to monitor the pulsar and up to the year 2001; a set of 3400 images has been built. It covers more than 20 hours of photometry data (exposure times of the images are between 4 and 15 seconds and sampling rates are between 4.0 and 26.7 seconds) taken with four different telescopes (the Hubble Space Telescope, the 1.22 m and 1.82 m telescopes of the Asiago Observatory and the 2.12 m telescope of the Guillermo Haro Observatory). The data span a period of almost nine years (Čadež et al. 2001).

3. Data Analysis and Results

The following Fourier analysis on our data is applied to confirm the observed 60 second modulation. For each obtained light curve, $m_k(t_l)$, we calculate its Fourier transform,

$$F_k(\omega) = \frac{1}{\sqrt{2\pi(t_{l_{max}} - t_{l_{min}})}} \sum_l m_k(t_l)\, e^{i\omega t_l}\, \frac{t_{l+1} - t_{l-1}}{2} \tag{1}$$

in order to test if all Fourier transforms contain a common signal at a given frequency. However, since the rotation frequency ω_{rot} of the pulsar is slowly decreasing one should expect that the free precession frequency ω_{fp} should also be decreasing. Therefore we have to allow for the recalibration of frequency scale to a common date in order to take this effect into account. The theoretical prediction (Čadež et al. 1997) is that ω_{fp} can change as ω_{rot}^p, where p can vary from 1 to 3. Thus, we rescale all the Fourier transforms to the same date (Jan. 1, 1996) as

$$\tilde{F}_k(\omega) = F_k\left(\omega\left(\frac{\omega_{rot}^{\circ}}{\omega_{rot}}\right)^p\right) , \tag{2}$$

where ω_{rot}° is the pulsar's rotation frequency at the reference date and ω_{rot} is the pulsar's rotation frequency at the time of observations. For the value of p we take 1 (Čadež et al. 2001).

With $\tilde{F}_k(\omega)$ we construct a matrix that has for its components cross-correlation functions between pairs of Fourier transforms

$$M_{jk}(\omega) = \tilde{F}_j(\omega)\tilde{F}_k(\omega) . \tag{3}$$

If m_k were statistically independent random processes, then $M_{jk}(\omega)$ for all j, k and given ω would be a random process distributed in the complex plane according to a $2D$ Gaussian distribution. If, however, $M_{jk}(\omega)$ contain a common signal, then they will be distributed in a ring around the origin of the complex plane. The probability that a common signal is present is large if the effective width of the ring is significantly smaller than its radius.

At a discrete sample of frequencies ω_s, where $\nu_s = \omega_s/2\pi$ goes from 0.007 Hz to 0.02 Hz in steps of 3.3×10^{-4} Hz, we calculate radial distributions $\frac{dN}{d|M_{jk}|}(\omega_s)$ of cross-correlation functions $M_{jk}(\omega_s)$. The number of cross-correlated pairs binned in 45 bins of width $3.3\times10^{-4}\mathrm{mag}^2\ \mathrm{Hz}^{-1}$ is 136 in the case of the Crab pulsar and 78 in the case of a test star. The results for the test star and the Crab pulsar are shown in Figure 2. In the case of the test star all distributions are similar which confirms the assumption of Gaussian noise. It is apparent that the Crab pulsar is noisier than the test star which is also in agreement with Čadež et al. (2001). At a frequency ~ 0.0167 Hz we can see brighter colors on the contour plot for the values of $|M_{jk}|$ between 0 and 0.004 mag^2 Hz^{-1} on one side and "a darker island" for the values of $|M_{jk}|$ around 0.007 mag^2 Hz^{-1}. It is clear that a periodic signal at the frequency $\sim 1/60$ Hz is present.

4. Conclusions

Our evidence suggests that the Crab pulsar free precesses. However, the measured free precession signal is too faint to give a final answer on whether

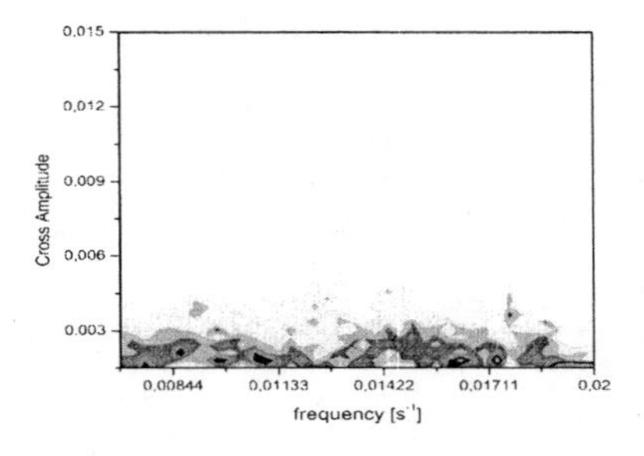

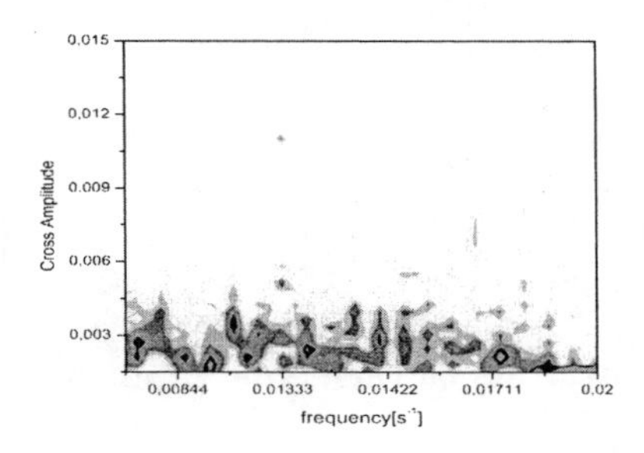

Figure 2. Contour plot representing radial distributions of cross-correlation functions as a function of frequency for the test star (left) and the Crab pulsar (right). The x-axis represents the recalibrated frequency in Hz and y-axis represents the amplitude of the cross-correlation function in mag^2 Hz^{-1}. Number of M_{jk} binned in 45 bins is presented with different shades of gray, the brighter is the color the lower is the number.

the amplitude of the free precession changes with time and on what time scale the pulsar relaxes its internal stress. Examining the free precession one can learn about the equation of state of the neutron matter in pulsars. Moreover, Chandra observations of the Crab pulsar and its nebula (Weisskopf et al. 2000) have given additional motivation to understand interaction mechanisms between the pulsar and highly magnetized plasma around it. We are continuing to observe the Crab pulsar in stroboscopic mode. The increase in signal to noise ratio of the free precession signature will help answer the remaining questions regarding pulsar physics.

References

Cocke, W.J., Disney, M.J. & Taylor, D.J. 1969, Nature, 221, 525

Čadež, A. & Galičič, M. 1996, A&A, 306, 443

Čadež, A., Galičič, M. & Calvani, M. 1997, A&A, 324, 1005

Čadež, A., Vidrih, S., Galičič M. & Carramiñana, A. 2001, A&A, 366, 930

Percival, J.W. et al. 1993, ApJ, 407, 276

Weisskopf, M.C. et al. 2000, ApJ, 536, L81

Part 2. Contributed Papers

Section B. Galaxies and Cosmology

International Symposium on Astrophysics Research and on the Dialogue between Science and Religion, Vatican Observatory, 2003
C. D. Impey & C. E. Petry, editors

Galaxies: The Link between Stars and the Universe

Vladimir Avila-Reese

Instituto de Astronomía, UNAM, A.P. 70-268, 04510 México, D.F.

Abstract. I review briefly the inflationary-motivated Cold Dark Matter cosmogony, and the formation and evolution of disk galaxies within the context of this model. I focus on results obtained by our group at UNAM.

1. From Quantum Fluctuations to Protogalaxies

Galaxies are a key link in the history of the Universe. On one hand they are the ecosystems where stars, planets, satellites, comets, etc. are born, live and die interacting with the interstellar medium. On the other hand, they are the structural unities of the Universe. According to the currently popular scenario of cosmic structure formation, the seeds of galaxies were imprinted in the first fraction of a second of the history of the Universe. It is postulated that at these remote epochs the Universe was filled by vacuum energy (or another kind of scalar field). Fluctuations at the scales of physical interest today could have been generated by microscopic quantum processes inherent to the vacuum. The wavelengths of these fluctuations were stretched exponentially, becoming supra-horizon perturbations to the space-time metric, which expanded into causally connected density perturbations. This process happened during an inflationary period triggered by instabilities of the scalar quantum field, which had an equation of state with negative pressure. Through a transition phase, the inflation ended with the generation of the "normal" hot matter and radiation in an expanding Universe. This is the basis of the *inflationary cosmology* (Guth 1981), which solves the limitations of the classical hot Big Bang theory.

The inflationary theory is in resonance with models of high energy physics that describe the breaking of internal gauge symmetries by which the elemental particles and interaction fields arise. In these models, several kinds of weak interacting particles are predicted which could be much more abundant than the normal baryonic particles. In fact, the observed abundances of light elements combined with the Big Bang nucleosynthesis show that the baryon density accounts for only $4 \pm 1\%$ of the critical density. On the other hand, studies of galaxy clustering on large scales, among other techniques, show that the total matter contribution to this density accounts for $30 \pm 5\%$. Therefore, around 85% of the matter in the Universe is not baryonic! Recent observational studies yield more surprises. Cosmological supernovae show that a repulsive form of energy dominates today the dynamics of the Universe. On the other hand, the sub-degree anisotropies of the CMB radiation reveal a flat, inflation-motivated, metric, which implies a critical value for the total density of the Universe. Therefore, roughly 70% of the Universe seems to be filled by the mysterious repulsive "dark energy", 26% by exotic dark matter, and only 4% by baryons. The existence of dark matter has also been revealed at galactic scales: to balance the centrifugal force of the rotating galaxy disks, large amounts of unseen matter must be invoked. Theoretical arguments are also advocated in favor of exotic dark matter. The seeds of galaxies would have been erased during the matter-radiation coupling epoch if their densities were dominated by baryons. However, only cold (and perhaps warm) non-baryonic dark matter particles guarantee the survival of perturbations on galactic scales. This is because cold dark matter (CDM) particles become non-relativistic at very early epochs and in this way the fluctuations do not suffer the free-streaming damping.

Since the primordial density perturbations produced during inflation are dominated in density by CDM, the main driver of their evolution is gravity. After some time, they collapse and form self-gravitating structures (CDM halos), where the gas further cools, condenses, and transforms into stars (White & Rees 1978; White & Frenk 1991). The amplitude-scale relation of the primordial perturbations is such that cosmic structures collapse from the smaller to the larger scales. The first step for galaxy formation in this *hierarchical CDM scenario* is the non-linear collapse of the dark halos. This task has been approached using numerical and semi-numerical simulations (e.g., Frenk et al. 1985; Navarro et al. 1997; Avila-Reese et al. 1998). The cosmological N-body simulations for the CDM model with dark energy (ΛCDM) have shown an impressive agreement with observations of large-scale structure in the Universe, where the physics of baryon matter is not relevant to the gravitational dynamics.

To study galaxy formation, the baryon matter processes must be included. One of the crucial questions about galaxies is which cosmological initial conditions and/or astrophysical processes are relevant in explaining their properties, correlations and evolutionary features, in particular the Hubble morphological sequence. Since disk galaxies are the dominant population of normal galaxies today, much effort has been made to understand these galaxies in the last few years.

2. Disk Galaxies: The Test-Bed of CDM at Small Scales

The formation of disks is a generic prediction of the CDM cosmogony because most dark halos are born with some initial angular momentum. The gas trapped within the virialized halos dissipates and collapses by large factors before settling into a disk in centrifugal equilibrium (Fall & Efstathiou 1980). The most direct approach to self-consistently model the processes of dark halo formation, gas dissipation, collapse into a disk, star formation and feedback is to use cosmological N-body plus hydrodynamical simulations. The community expects in the next years fully self-consistent results from these simulations. In the meantime, some workers have developed alternative semi-numerical approaches with exciting results, which would allow the interpretation of several observational facts. Below we present briefly our approach and some key results (Avila-Reese & Firmani 2000; Firmani & Avila-Reese 2000).

Star formation (SF) is a key process in galaxy evolution. Firmani & Tutukov (1992) introduced a model of disk large-scale SF, where SF is triggered by gravitational instabilities and self-regulated by a balance between the SN energy input and the energy dissipated due to turbulent motions in the ISM. The model predicts the observed dependence of the SF rate on gas density and shows that disk galaxies would consume their gas in $t \lesssim 3-5$ Gyr. Thus, gas infall is necessary. In the context of the CDM scenario, galaxies form through a continuous process of mass aggregation. Assuming that smooth accretion dominates over mergers and that gas collapses with detailed angular momentum conservation, we modeled the formation of disks in centrifugal equilibrium within growing CDM halos. Thus, the SF history of the galaxy models is driven by the cosmological gas infall rate, and the disk gas surface density is determined mainly by the angular momentum of the halo. The former assumption is made because disks are dynamically fragile objects and collisions would heat and thicken the stellar disk beyond the level observed at present; cosmological simulations show indeed that most of galaxy-sized halos grow quiescently. The latter assumption is arguable and currently is among the key questions of disk formation models (see e.g., van den Bosch et al. 2002). In our model, bulges form from secular evolution of the stellar disk.

The model results for a ΛCDM cosmology agree with a large body of observational data. We conclude that the main properties of disk galaxies (exponential surface brightness profiles, nearly flat rotation curves, local and global color indexes, gas fractions f_g, etc.) as well as their correlations (Tully-Fisher and magnitude-radius relations, the Hubble sequence, etc.) are the result of the combination of three cosmological factors and their dispersions: the total mass, the mass aggregation history (MAH) and λ. The last two determine the intensive properties and are at the basis of the Hubble sequence correlations. The MAH influences the integral colors, f_g, and halo structure, while λ influences the surface brightness, b/d ratio and rotation curve shape. The higher the surface brightness and the redder the disk, the larger the b/d ratio and the lower f_g. The mass has almost no influence on these properties, the Hubble sequence being most likely a biparametrical sequence driven by the MAH and λ. The predicted Hubble sequence correlations are less scattered than observations. Processes of secondary importance for disk galaxies, such as minor mergers and non-stationary SF, could produce more spread around these correlations.

The empirical Tully-Fisher relation is a direct consequence of the mass-velocity relation of the CDM halos, determined mainly by the power spectrum of the primeval fluctuations. The models predict that the Tully-Fisher of low and high SB galaxies is roughly the same, as observed. The residuals of the Tully-Fisher relation and their correlation with galaxy properties also agree with observations. Models can be tested even in the local Universe. For a Milky Way fiducial model, the predicted evolution of the SF rate in the solar neighborhood agrees with the evolution derived from the H-R diagram provided by the Hipparcos satellite (Hernández et al. 2001).

The models have a free parameter, f_d, the disk-to-total mass ratio. To reproduce jointly the observed shapes of rotation curves and the galaxy surface brightness distribution, f_d should be fixed to $\approx 3\%$ of the total halo mass. This is much less than the universal baryon-to-total mass ratio ($\sim 14\%$). Normal matter is only $\sim 4\%$ of all matter-energy in the Universe, and only $\sim 20\%$ of this fraction is actually in galaxies and stars! Detailed studies on the halo gas dissipation and feedback efficiencies indeed showed that only $30 - 50\%$ of the halo baryon fraction ends in the disks (van den Bosch 2002). On the other hand, large-scale cosmological simulations including gas show that only $30 - 40\%$ of the universal baryon fraction ends within collapsed halos today; the rest is in the form of warm-hot intergalactic gas in the filaments and voids (Davé et al. 2001).

A potential shortcoming of the CDM scenario is the high concentration it predicts for the galactic halo cores. We have re-simulated our models allowing for the formation of shallow cores in the CDM halos so that they resemble the cores of LSB and dwarf galaxy halos. The shapes of rotation curves, the maximum disk-to-total circular velocity ratios, and the zero-point of the Tully-Fisher relation moved into better agreement with the observational data. Realistic shallow cores can be produced introducing self-interaction in the CDM particles (Colín et al. 2002). By observing galaxies, one can infer properties of the exotic elementary particles. This is the beauty of astronomy!

3. What Next?

According to our models, the global SF history produced by the disk galaxy population agrees with observational inferences back to $\sim 8 - 10$ Gyr ago (the age of the Universe is ~ 14.5 Gyr). At earlier epochs, the observed global SF rate continues to increase (Lanzetta et al. 2002). The SF and luminosity at these epochs could be produced by elliptical galaxies and bulges of spirals. While the evolution of disks appears gradual and extended in time, spheroids seem to concentrate their SF activity in a violent, early phase of their evolution. Nowadays, the increasing information on the high-redshift universe is opening new perspectives in the understanding of the formation of spheroids, its connection with QSOs, ULIRGs, LBGs and other high redshift objects, as well as in the interplay of galaxy formation in general with the intergalactic medium. A better understanding of the disk assembling, hydrodynamics and SF processes of disk galaxies, as well as a self-consistent model for the formation of spheroids, are among the challenges for the coming years.

Galaxies are fossil records of the early Universe and its history. Models of formation and evolution of disk galaxies within the context of the most popular cosmology, and for the now accurately determined cosmological parameters, have been shown to be in agreement with observations. If somebody asks if the existence of God been revealed in this study, the answer is no. But I would add that this is not the correct place where to look for God since then demonstration of the existence of God is not the scope of science. If anything, science may show us what is not God; and following George Coyne, I would say that this frees us from one more potential kind of idolatry.

Acknowledgments. I am very grateful to the organizers, to the Vatican Observatory Staff, and to all the participants of this interesting and friendly meeting. My research is supported by CONACyT grant 33776–E.

References

Avila-Reese, V., Firmani, C. & Hernández, X. 1998, ApJ, 505, 37
Avila-Reese, V. & Firmani, C. 2000, Rev. Mex. A&A, 36, 23
Colín, P. et al. 2002, ApJ, 777, 581
Davé et al. 2001, ApJ, 552, 473
Fall, S.M. & Efstathiou, G. 1980, MNRAS, 193, 189
Firmani, C. & Avila-Reese, V. 2000, MNRAS, 315, 457
Firmani, C. & Tutukov, A.V. 1992, A&A, 264, 37
Frenk, C.S. et al. 1985, Nature, 317, 595
Guth, A.H. 1981, Phys. Rev. D, 23, 347
Hernández, X., Avila-Reese, V. & Firmani, C. 2001, MNRAS, 327, 329
Lanzetta, K.M. et al. 2002, ApJ, 570, 492
Navarro, J.F., Frenk, C.S. & White, S.D.M. 1997, ApJ, 490, 493
van den Bosch, F. 2002, MNRAS, 332, 456
van den Bosch, F. et al. 2002, ApJ, 576, 21
White, S.D.M. & Rees, M. 1978, MNRAS, 183, 341
White, S.D.M. & Frenk, C.S. 1991, ApJ, 379, 52

International Symposium on Astrophysics Research and on the Dialogue between Science and Religion, Vatican Observatory, 2003
C. D. Impey & C. E. Petry, editors

Semi-Analytic Model of the Intra-Cluster Medium Metal Enrichment

Sofía A. Cora and Simon D.M. White
Max-Planck-Institut für Astrophysik, Karl-Schwarzschild-Strasse 1, 85740, Garching bei München, Germany

Abstract. The metal enrichment of the intra-cluster medium (ICM) is investigated using a method that combines N-body simulations and a semi-analytic model of galaxy formation. The cluster of galaxies is simulated in a flat, low density universe, with a numerical resolution that allows the detection of substructures in the dark matter background of the cluster. The phenomenological approach used to model the physical processes involved in the galaxy formation and metal production is applied to the substructures found in the dark matter halos detected at different redshifts. Details of the chemical implementation in the semi-analytic model and first results related to the mean properties of the baryonic matter components are presented.

1. Introduction

A new generation of X-ray observatories provides a great deal of information on the thermal and chemical properties of the hot gas trapped in the potential well of galaxy clusters. These structures are the largest virialized ones and are good approximations to "closed boxes," thus being suitable for the study of processes involved in the chemical enrichment of their baryonic components. The detection of Fe K line emission was the most significant observational discovery concerning X-ray clusters. The content of iron in the intra-cluster medium (ICM) supports the fact that some of its hot gas originates in galaxies that reside in the cluster. The physical mechanisms considered for the supply of this enriched gas include galactic winds driven by supernovae (SNe) explosions (White 1991; Renzini 1997), ram pressure stripping (Mori & Burkert 2000), and early enrichment by hypernovae associated with population type III stars (Loewenstein 2001). Observations from satellites like HEAO 1, EXOSAR, Ginga and ROSAT indicate that the iron abundances range from 25% to 50% of the solar value. Abundance gradients in the ICM of some clusters have also been measured (White 2000; Finoguenov et al. 2000; De Grandi & Molendi 2001), providing clues about the relative contribution of different type of supernovae (SNe) and about the mechanisms that determine the presence (or absence) of metallicity profiles in cooling flow (or non-cooling flow) clusters. These data put constraints on the origin of metals and help to understand the physical processes involved in the formation and evolution of galaxies. However, their interpretation can lead to controversial results, and complementary numerical studies are necessary to disentangle the numerous puzzles related to the metal enrichment of the ICM.

2. Hybrid Model of ICM Metal Enrichment

In a hierarchical structure formation scenario, large scale structure is formed by the clustering of smaller entities. In order to estimate the impact of the hierarchical build up of structure on the chemical enrichment of the ICM, we used a method that combines cosmological N-body simulations of clusters of galaxies, and a semi-analytic model (SAM) of galaxy formation (Springel et al. 2001).

We employed the GIF-ΛCDM model with cosmological parameters ω_o=0.3, ω_Λ=0.7, $H_0 = 100\,h\,\mathrm{km\,s^{-1}\,Mpc^{-1}}$, with $h = 0.7$, spectral shape $\Gamma = 0.21$, and cluster-nomalization of $\sigma_8 = 0.9$. The second most massive cluster in this simulation ($M_{\mathrm{vir}} = 8.4 \times 10^{14}\,h^{-1}\,\mathrm{M}_\odot$) was resimulated with higher resolution in the region immediately surrounding the object with $\approx 2.2 \times 10^5$ dark matter particles of mass $1.4 \times 10^{10}\,h^{-1}\,\mathrm{M}_\odot$. The simulation was carried out with the parallel tree code GADGET.

The partially stripped haloes of cluster galaxies can be identified within the dark matter clumps corresponding to groups and clusters when sufficiently high mass resolution simulations are used. All these substructures were found by using the SUBFIND algorithm and were used by the SAM to generate the population of galaxies (Springel et al. 2001). This semi-analytic model was modified to take into account the circulation of metals among the different baryonic components, providing the mean properties of the galaxies and the ICM. Metal production prescriptions were added to the recipes of hydrodynamics of cooling, star formation, supernovae feedback and galaxy-galaxy merging. Stars can contaminate the cold and hot gas because of episodes of mass loss during their evolution and metals ejected at the end of their lives. In the present version of the SAM, only the latter mechanism is considered, by the inclusion of effects of supernovae type II (SNe II) and type Ia (SNe Ia). The temporal behaviour of the abundances of several chemical elements is followed, instead of adopting the yield as a free parameter as in other semi-analytic models (Kauffmann & Charlot 1998; Cole et al. 2000; Somerville et al. 2001). The metal production is estimated for each star formation episode. For SNe II, the yield of metals (Woosley & Weaver 1995) is assumed to be immediately recycled to the cold gas, allowing some fraction to go directly to the hot halo gas (Kauffmann & Charlot 1998; Cole et al. 2000). The number of SNIa is obtained by assuming a scaling of the rate between the two types of SNe based on observed values (Tsujimoto et al. 1995: SNRIa/SNRII = 0.15–0.3). The SNIa ejecta (Iwamoto et al. 1999) are assumed to be produced at a given time after a starburst (0.1–1 Gyr). Starbursts induced in major mergers contribute metals that contaminate the hot gas since the cold gas is assumed to be completely consumed in this process. We consider the metallicity dependence of the cooling rate and of the population synthesis models used to calculate the photometric properties of the galaxies (Kauffmann & Charlot 1998).

3. Results

The model was normalized following Kauffmann et al. (1999). The parameters introduced were tuned to match observed metallicity properties of the galaxies and the cluster. The metallicity of the stellar population of "Milky Way" like galaxies can be reconciled with the values observed in the stars of the solar neighborhood (mean metallicity of 0.7 times the solar value) by increasing the yield of heavy elements by a factor of two, taking into account the theoretical uncertainties in the nucleosynthesis calculations (Gibson et al. 1997). However, the metallicity of the hot gas in the cluster could not be reproduced. The value obtained is $\simeq 50\%$ lower than the observed values. We can understand the origin of this lack of metals in the ICM by considering mean properties of the cluster. With the parameters adopted, we obtained a mass of hot gas $M_{\rm ICM} = 1.47\times10^{14}{\rm M}_\odot$ and a total stellar mass of the galaxies contained in the cluster $M_\star = 8.27\times10^{12}{\rm M}_\odot$. Comparing these values with those adopted by White et al. (1993) for the Coma Cluster ($M_{\rm ICM} \simeq 5.5\times10^{13}\,h^{-5/2} = 1.34\times10^{14}{\rm M}_\odot$, $M_\star \simeq 10^{13}\,h^{-1} = 1.43\times10^{13}{\rm M}_\odot$), we found that the stellar mass in the simulated cluster is a factor 60% of the observed mass. This is consistent with the behaviour shown by the B-band luminosity function constructed from the galaxies in the cluster, which only matches the bright end of the composite luminosity function of galaxy clusters of Trentham (1998). Hence, a significant fraction of galaxies is missing in the simulated cluster, which corresponds to stellar mass produced in haloes that are below the resolution limit of the simulation. This leads to low mass of metals ejected into the ICM.

A relevant indicator of the mean properties of the cluster is the ICM iron mass to light ratio, $M_{\rm ICM}^{\rm Fe}/L_{\rm B}$, where $M_{\rm ICM}^{\rm Fe}$ is the iron content in the ICM, and $L_{\rm B}$ is the optical luminosity of galaxies in the cluster. The observed value of this ratio ranges within 0.01–0.02 $M_\odot/L_\odot$ (Renzini 1997). For the simulated cluster, we found $M_{\rm ICM}^{\rm Fe}/L_{\rm B} = 0.019\,M_\odot/L_\odot$, in good agreement with observations. This is an important result since it shows that the circulation of metals between the different baryonic components (stars and gas) in the cluster is being well tracked.

These results call for the analysis of higher resolution simulations that are already available, which also follow the dynamical evolution of gas particles that represent the ICM using smoothed particle hydrodynamics.

The analysis of the chemical enrichment of these gas particles will allow us to investigate the processes that give rise to the final spatial distribution of metals in the ICM and its evolution with redshift.

Acknowledgments. We thank N. Yoshida for making their simulations available for the present study, and V. Springel and F. Stoehr for their help and useful discussion. S.A. Cora is grateful for the hospitality of Max-Planck Institut while doing this research. This work was supported by a grant from Fundación Antorchas.

References

Cole, S., Lacey, C.G., Baugh, C.M. & Frenk, C. 2000, MNRAS, 319, 168

De Grandi, S. & Molendi, S. 2001, ApJ, 551, 153

Finoguenov, A., David, L.P. & Ponman, T.J. 2000, ApJ, 544, 188

Gibson, B.K., Loewenstein, M. & Mushotzky, R.F. 1997, MNRAS, 290, 623

Iwamoto, K., Brachwitz, F., Nomoto, K., Kishimoto, N., Umeda, H., Hix, W.R. & Thielemann, F.-K. 1999, ApJS, 125, 439

Kauffmann, G. & Charlot, S. 1998, MNRAS, 294, 705

Kauffmann, G., Colberg, J.M., Diaferio, A. & White, S.D.M. 1999, MNRAS, 303, 188

Loewenstein, M. 2001, ApJ, 557, 573

Mori, M. & Burkert, A. 2000, ApJ, 538, 559

Renzini, A. 1997, ApJ, 488, 35

Somerville, R.S., Primack, J.R. & Faber, S.M. 2001, MNRAS, 320, 504

Springel, V., White, S.D.M., Tormen, G. & Kauffmann, G. 2001, MNRAS, 238, 726

Trentham, N. 1998, MNRAS, 294, 193

Tsujimoto, T., Nomoto, K., Yoshii, Y., Hashimoto, M., Yanagida, S. & Thielemann, F.-K. 1995, MNRAS, 277, 945

White, R. 1991, ApJ, 367, 69

White, S.D.M., Navarro, J.F., Evrard, A.E. & Frenk, C.S. 1993, Nature, 366, 429

White, D.A. 2000, MNRAS, 312, 663

Woosley, S.E. & Weaver, T.A. 1995, ApJS, 101, 181

International Symposium on Astrophysics Research and on the Dialogue between Science and Religion, Vatican Observatory, 2003
C. D. Impey & C. E. Petry, editors

A Search for Quasars behind Nearby Galaxies

A. Dobrzycki, P.J. Groot[1], L.M. Macri[2] and K.Z. Stanek

Harvard-Smithsonian Center for Astrophysics, 60 Garden Street, Cambridge, MA 02138, USA

Abstract. Quasars behind nearby galaxies are of great astronomical interest, but until recently they were rather hard to identify. Superb spatial resolution of the *Chandra X-Ray Observatory* and availability of data from optical monitoring campaigns have recently led to discoveries of several quasars behind the Magellanic Clouds. We present results of our ongoing projects for the identification of such quasars. We discuss the prospects for searches for quasars behind other dense stellar fields.

1. Introduction

Searches for quasars in dense stellar fields—such as the Magellanic Clouds—were in the past hampered by difficulties in selecting candidates. Optical follow-up of X-ray selected objects had to deal with the large number of candidates in X-ray source error boxes, while variability studies required the monitoring of a vast number of objects. At the same time, such quasars are of great astrophysical interest, for example as reference points for analysis of proper motions and as background sources for absorption studies. Until recently, only a handful of confirmed quasars behind the LMC and SMC were known, and almost all of them were located behind the outer, sparse parts of the Clouds.

Recent developments—such as the launch of the *Chandra X-Ray Observatory* with its superb spatial resolution in X-rays and the availability of large photometry databases (OGLE, MACHO)—are now making systematic searches for quasars in dense stellar fields possible. Those developments led to discoveries of several quasars behind the LMC and SMC.

---

[1]Present address: Department of Astrophysics, University of Nijmegen, PO Box 9010, 6500 GL Nijmegen, The Netherlands

[2]Present address: National Optical Astronomy Observatory, 950 North Cherry Avenue, Tucson, AZ 85726, USA

In this contribution, we present recent results from our project that utilizes both X-ray- and optical variability-based selection of quasar candidates.

2. Quasars behind the Magellanic Clouds

2.1. X-Ray-Selected Quasars

A good starting point for searching for quasars is to utilize the fact that they are often bright in X-rays, and a considerable number of quasars have been found in that way (e.g. Boyle et al. 1993). However, the limited spatial resolution of earlier X-ray missions rendered this method impractical in dense stellar fields, where a single X-ray source would have many optical candidates in its error box. The launch of the *Chandra X-Ray Observatory* made application of this method to the Magellanic Clouds possible. *Chandra*'s superb spatial resolution and excellent positional accuracy significantly reduce the source confusion problem.

We are pursuing a project of matching *Chandra* serendipitous sources with objects in the OGLE database. While *Chandra* has an excellent positional accuracy, it has a very spotty sky coverage, since it only performs pointed observations. At the time of this writing, only five *Chandra* pointings towards the LMC and two pointings towards the SMC overlap with the OGLE fields.

We retrieved the observations from the *Chandra* archive and ran source detection programs on them. To prioritize our follow-up observations, and whenever it was practical, we performed X-ray spectral analysis and identified sources with absorbed and/or hard spectra. We completed the candidate lists by identifying the corresponding sources in the OGLE database. The follow-up observations were performed with the Magellan Baade 6.5-meter telescope in January and September of 2002.

As of the time of this writing, we have identified four quasars behind the LMC in effectively 0.15 deg^2 and seven behind the SMC in 0.5 deg^2 (Dobrzycki et al. 2002a; 2003, in preparation). All but one of those quasars lie behind very dense stellar fields in the LMC and SMC. They are thus excellent objects to be used as reference frames for the studies of proper motions and for the analysis of absorption in the Clouds.

2.2. Variability-selected Quasars

Another characteristic that was often used in quasar surveys was their irregular variability. Several quasar surveys utilizing variability were performed or are on-going (e.g. Meusinger & Brunzendorf 2001; Rengstorf et al. 2001). In the Magellanic Clouds, this technique has been used by Geha et al. (2002), who performed spectroscopic follow-up observations of $\sim$260 candidates selected from the MACHO database. They found thirty-eight

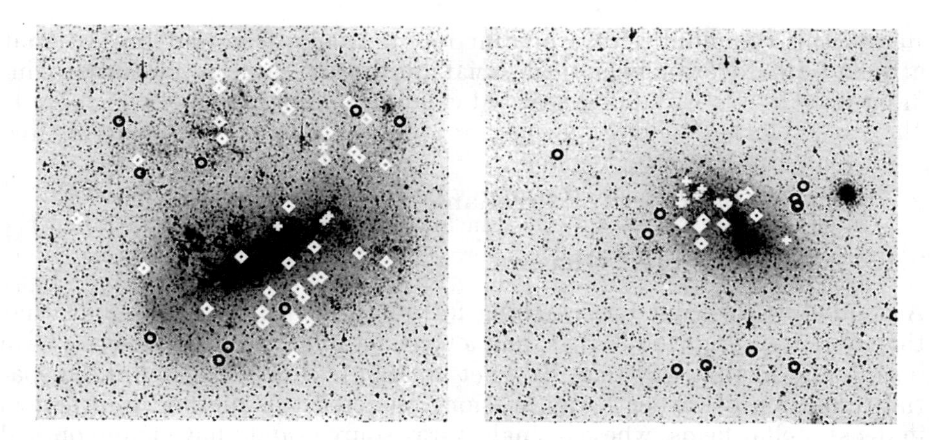

Figure 1. Quasars in the direction of the LMC (left panel) and SMC (right panel). Black circles indicate quasars known before early 2002. White diamonds show quasars identified by the MACHO collaboration (Geha et al. 2002). White crosses show quasars identified by us (Dobrzycki et al. 2002a, 2002b, 2003, in preparation). Images courtesy of G. Bothun.

quasars in 35 deg^2 behind the LMC and nine quasars in 2.5 deg^2 behind the SMC.

Eyer (2002) presented an algorithm for selecting quasar candidates from objects in the OGLE database. He searched the database for slowly and irregularly varying blue objects and identified QSO candidates towards the Magellanic Clouds: 118 towards the LMC and 15 towards the SMC. In 2002 September, we performed follow-up spectroscopical observations of the twelve brightest of Eyer's candidates, and identified five of them as quasars (Dobrzycki et al. 2002b). Two of those quasars were also identified by Geha et al.

Eyer's and Geha et al.'s candidate lists have a rather small overlap (only $\sim$10 objects in common). This fact and the fact that X-ray-selected quasars are typically redder than the variability-selected objects indicate that both techniques are conservative and both miss some quasars. The two methods should therefore be considered as complementing one another, rather than as competing.

3. Discussion

In Figure 1 we show the locations of known quasars in the direction of the Magellanic Clouds, both "old" and those recently discovered. We stress that in this case "old" refers to early 2002, only a few months ago at the time of this writing. A qualitative difference in the coverage and large increase in the number of available objects are clearly visible. We note, however, that quasars behind the LMC bar are still underrepresented. This is because Eyer's variability selected candidates, almost all of which are in the bar, were not followed up yet, and because Geha et al. concentrated their search in the outer regions.

Another extremely interesting region to search for quasars is the Galactic Bulge. To our knowledge, no quasars have been identified so far in the vicinity of the Galactic center. There are, however, several regions where interstellar extinction is quite low (such as Baade's Window) and where one could, in principle, see quasars.

At present, there are no good *Chandra* pointings in the direction of the low extinction windows. However, variability-based search is promising. Identifying quasars behind the Galactic Bulge would be very valuable. Recently, Sumi, Eyer, & Woźniak (2002) have shown that there is a statistically significant difference between proper motions of faint versus bright red clump stars in one of the OGLE bulge fields. Finding fixed reference points for this study, which quasars could provide, would be an extremely interesting result.

On one hand, the search toward the Galactic bulge will be made easier by the fact that there will be few early type stars, which in the Magellanic Clouds are the primary contaminants in the candidate lists. On the other hand, the quasars will be considerably reddened even in the low extinction windows, making them less conspicuous as blue objects. Also, the surface density of objects that need to be analyzed will be very large, and as a result there will likely be a sizeable number of artifacts, etc., which will contaminate the candidate lists.

Acknowledgments. AD acknowledges support from NASA Contract No. NAS8-39073 (CXC).

References

Boyle, B.J., Griffiths, R.E., Shanks, T., Stewart, G.C. & Georgantopoulos, I. 1993, MNRAS, 260, 49

Dobrzycki, A., Groot, P.J., Macri, L.M. & Stanek, K.Z. 2002a, ApJ, 569, L15

Dobrzycki, A., Macri, L.M., Stanek, K.Z. & Groot, P.J. 2002b, AJ, in press (astro-ph/0210601)

Eyer, L. 2002, AcA, 52, 241

Geha, M. et al. 2002, AJ, in press (astro-ph/0209513)

Meusinger, H. & Brunzendorf, J. 2001, A&A, 374, 878

Rengstorf, A.W. et al. 2001, AAS Mtg., 199, 138.07

Sumi, T., Eyer, L. & Woźniak, P.R. 2002, MNRAS, in press (astro-ph/0210381)

International Symposium on Astrophysics Research and on the Dialogue between Science and Religion, Vatican Observatory, 2003
C. D. Impey & C. E. Petry, editors

Galaxy Quest... for Sense

Jose Funes, S.J.

Vatican Observatory, University of Arizona, Tucson, AZ 85721, USA

Abstract. In this work I explore the impact of astrophysical research, particularly the field of extragalactic research, on our everyday culture and, especially, on our search for meaning.

1. Galaxies in Our World

The word "galaxy" has become part of our daily vocabulary. It has been used in different contexts. For instance, in the classical science-fiction movie "Star Wars," the story begins by saying: "A long time ago, in a galaxy far, far away...." However the term "galaxy" is not only restricted to the science-fiction world or to the science-reality world; some years ago an Italian newspaper referred to the political left-center coalition as the "galaxy of the left-center." Moreover "Galaxy" is also the name of a car model. As we see, galaxies have entered our popular world. We might say that the concept of "galaxy" is the "Copernican revolution" of the twentieth century.

Although the philosopher Immanuel Kant proposed in 1775 that "spiral nebulae" were analogies of the Milky Way, only in 1920 with the "Great Debate" did the astronomical community started to shed light on the nature of the "spiral nebulae" so that they could find their place within the cosmic order. During the debate, Shapley and Curtis argued over the scale of the universe. Curtis argued that the Universe is composed of many galaxies like our own. Shapley argued that the "spiral nebulae" were really nearby gas clouds, and that the Universe was composed of only one big Galaxy. A partial resolution of the debate came in 1924. Using the 100-inch Hooker Telescope at Mount Wilson, then the largest telescope in the world, Edwin Hubble was able to prove that Andromeda was another galaxy, remote from the Milky Way. In synthesis, in the twentieth century we have learnt that we live in a disk galaxy and we have also learnt to locate our place in the known universe. It seems that humanity does not have a privileged position in the universe. We live on the third planet of a stellar system among billions of stars in a galaxy that is only one among roughly 100 billion galaxies.

In the 1990s, Hubble Space Telescope images, such those taken in the Hubble Deep Field, help us to realize that we are part of a galaxy in an universe of 100 billion galaxies. Astronomers that study galaxies can repeat with genuine amazement Isaac Newton's words:

> I see myself as a child play on the seaside and sometimes having fun discovering a small rock or a shell that is more beautiful than normal, while in front of me, the huge ocean of truth extends unexplored.

In front of this "huge ocean" called the universe, astronomers may ask themselves questions like "Why there is a universe at all?" or "Why are galaxies so beautiful?"

The title of this paper is inspired by the movie "Galaxy Quest," not because the movie itself has a special value or interest to me, but what I try to express in the title, and to explore in this work is the impact of astrophysical research, particularly the field of extragalactic research, on our every-day culture, and specifically, on our search for meaning.

2. The Quest for Intelligibility and Meaning

Most of the following discussion is based on Peacocke (1993). In our relation to the world there are two kinds of searches, the search for intelligibility and the search for meaning.

In the first kind of search, the scientist tries to find the "best explanation" for the physical phenomena he observes. This quest is aimed at finding answers to questions like "Why do we live in a disk galaxy rather than in a elliptical one?". Science provides an answer by depicting the realities of the natural world in metaphorical language. For example, the hierarchical model gives a general picture of the process of galaxy formation.

In the second search, the process of comprehending the universe invokes another sort of "Why?" question. Questions such as "What is the explanation and meaning of a universe that generates from within itself, beings who seek truth, create beauty, strive for goodness, have moral purposes, and create community?" are motivated by the religious and philosophical quest. However this quest has its roots in the intellectual excitement of the search for the ultimate nature of the universe and in the awareness of the incompleteness of our real understanding of the cosmos that astronomers experience in their search for intelligibility in the universe.

3. A Transcendent Source of Meaning: God As Response

One answer to the religious quest is the Christian response: "For the Christian, God, as the power making for intelligibility, beauty and righteousness, may be said to explain the universe in that he gives it meaning and intelligibility, provides purpose and significance, and so sets all things within an overall context" (Ward 1974). In other words, we may say that God is the best model that fits the existential data. Even when this could seem a proper response, a "reasonable" response, I think that the "best explanation" is not enough to be a source of meaning for the religious quest. Many times a rational proof of the existence of God has led people to become unbelievers, obtaining the contrary effect to what was desired by the rational proof of God's existence.

I am not saying that to believe in God is irrational. What I try to stress is that a "good explanation" of the universe is not enough to believe in God. To believe in God we need some kind of "love" experience, not only a "rational one." In fact, the faith of the Church derives from its experience of God not from a "good model," a model that fits the existential data. The experiences and encounters with God in the Bible are the paradigm for all Christian believers. From biblical studies we have learnt that first, God is experienced as savior, and in a second moment, as creator (best model for explanation and meaning). Only in the second moment do we reflect and see that God, the creator, is a source of meaning.

I find that this kind of reasoning and questioning is worthy for everybody, not only for Christians. The quest for sense is a human quest and the search of meaning can help us to find ways of improving the quality of our lives. Perhaps we may find some way to integrate our spiritual lives, or in other words our beliefs with our work, or our research. Here are some questions that perhaps deserve an answer, or are simply worthy of being asked:

- Is there any relation between the search for a scientific explanation of the universe and the search for a source of existential meaning?
- What is the impact of the scientific research on my life?
- What is the impact of my beliefs on my scientific research, or on my professional relations?

References

Peacocke, A. 1993, Theology for A Scientific Age (SCMD Press: London), pp. 87-98

Ward, K. 1974, The Concept of God (Blackwell: Oxford), p. 148

International Symposium on Astrophysics Research and on the Dialogue between Science and Religion, Vatican Observatory, 2003
C. D. Impey & C. E. Petry, editors

Gas Inflow in Spiral Galaxies from Kiloparsec to Parsec Scales

Witold Maciejewski

Osservatorio Astrof. di Arcetri, L.go Fermi 5, 50125 Firenze, Italy, and Obserwatorium Astronomiczne Uniw. Jagiellońskiego, Poland

Abstract. I briefly review models and observations of gas inflow in spiral galaxies from the scale of kiloparsecs towards the central parsec. Asymmetries in galactic potentials are crucial in triggering such inflow, and strong inflow has been indeed observed in stellar bars. Nested bars have been proposed to trigger a cascade of inflow, but it can be shown that inner bars cannot funnel gas as the outer bars do. For certain potential and gas parameters, gas falls towards the centre in a spiral shock.

1. Introduction

Nuclear starbursts and Active Galactic Nuclei (AGN) have to be fed by mass transport into the active regions. Due to dissipation, gas inflow is easiest to trigger. There are two obvious dynamical mechanisms causing gas inflow in galaxies: interactions and asymmetries in galactic potentials. The first one involves violent and transient phenomena, while the other can occur steadily in isolated galaxies, providing continuous feeding to the starburst/AGN activity. Here I focus on this second mechanism, and on the role of galactic bars in trigerring the inflow. Then I zoom in on the inner parts of galaxies, where the potential becomes axisymmetric, or where it is dominated by an inner, secondary bar.

2. Stellar Bars

Galactic discs commonly develop stellar bars, and the majority of observed galaxies is barred (e.g. Eskridge et al. 2000). The most established observational notion about gas morphology in barred galaxies is the existence of two straight or curved dust lanes in the main body of the stellar bar, which are symmetric around the galaxy centre, and slightly tilted to the bar's major axis (see e.g., NGC 1097, NGC 1300, NGC 4303, and NGC 6951 in the Hubble Atlas of Galaxies, Sandage 1961). Strong discontinuities in gas velocity, indicative of shocks, have been measured along the dust lanes (Regan et al. 1997). This morphology and the associated kinematics have been seen in numerical models of gas flow in a fixed potential of the bar since the work of Sørensen et al. (1976). It has been explained in terms of stable periodic orbits by Athanassoula (1992).

There are two major families of stable prograde periodic orbits inside the bar: the x_1 orbits parallel to the bar, and the x_2 orbits orthogonal to it (Contopoulos & Papayannopoulos 1980). The x_2 orbits are present only in galaxies with sufficient central mass concentrations and moderate bar pattern speeds—in linear analysis this condition corresponds to the existence of an Inner Lindblad Resonance (ILR, see Binney & Tremaine 1987). By definition, the x_2 orbits exist only inside this resonance. Representative orbits of each family are plotted in Figure 1.

3. Gas Flows

In the outer parts of the bar, gas tends to follow the x_1 orbits. Further inwards, the x_1 orbits develop cusps and loops: stars can follow such orbits, but gas on self-intersecting orbits gets shocked. A robust feature in the majority of numerical models is a pair of principal shocks that develop on the leading side of the bar (marked "PS" in Figure 1). There, torques transfer angular momentum from the gas to the bar, and the gas falls towards the centre. Numerical models indicate considerable gas inflow in the principal shocks: up to 1 $M_\odot\,yr^{-1}$ for typical gas densities in the disc (Athanassoula 1992, Regan et al. 1997).

The shocked gas usually does not flow straight to the galaxy centre: it tends to settle on the x_2 orbits, which have lower value of the Jacobi integral E_J (corresponding to the energy in the rotating frame). Consequently, the straight shocks are slightly tilted to the bar's major axis. If there were no x_2 orbits, straight shocks would develop on the bar's major axis, possibly extending directly to the centre of the galaxy (Athanassoula 1992, Piner et al. 1995). Only a couple of galaxies with such central (on-axis) straight shocks have been observed (Athanassoula 1992), which suggests that most barred galaxies possess the ILR. The gas settling on x_2 orbits, well inside the ILR, often accumulates in nuclear rings (marked "NR" in Figure 1), where high density and no shear make favorable conditions for star formation (Piner et al. 1995). Star-forming nuclear rings have been observed in many galaxies (e.g. NGC 1512, Maoz et al. 2001; NGC 4314, Benedict et al. 2002).

Not all barred galaxies possess nuclear rings though, because patterns of gas flow depend on the velocity dispersion in the gas. When it is low, then gas settles on the nuclear ring, but when it is high, then gas forms a nuclear spiral (Maciejewski 1998). This spiral is formed by the principal shock, which curls inwards at small radii, so that the curling shock from one side of the bar never meets its counterpart from the opposite side, and both of them propagate towards the centre as an $m = 2$ spiral mode (Figure 1, insert). The convergence of the flow in such created nuclear spiral is high

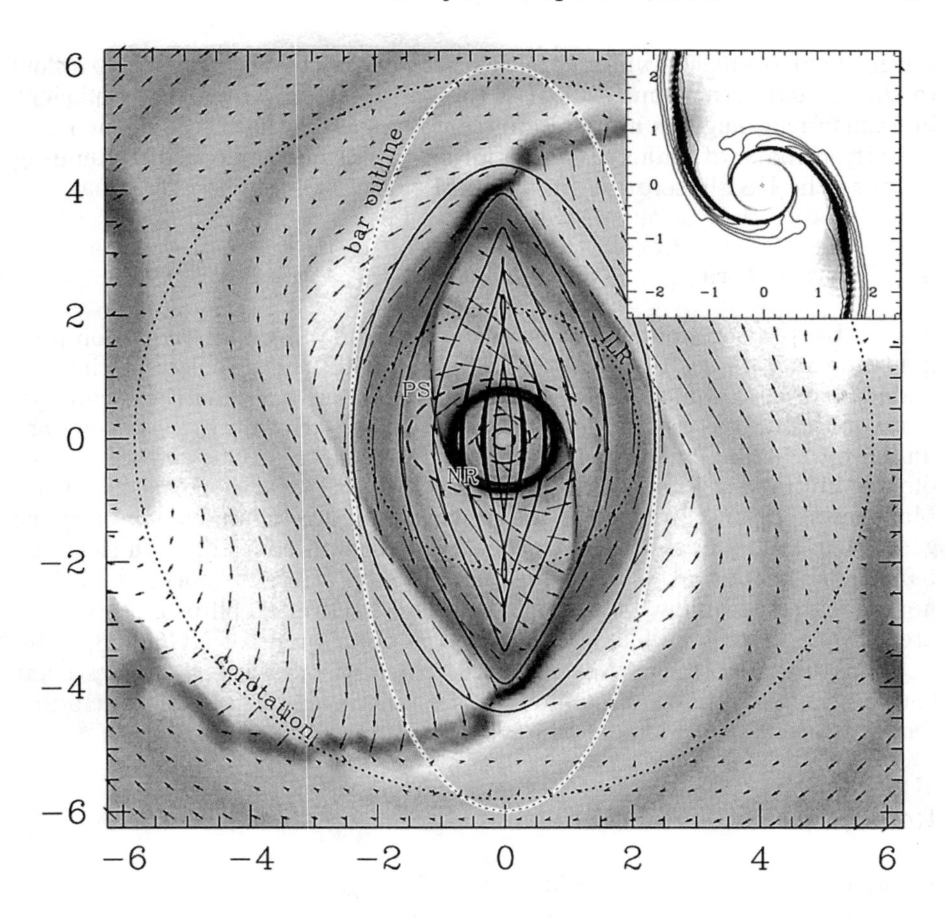

Figure 1. A representative snapshot of the gas density and the velocity field in a barred galaxy. The gas, treated as a non-self-gravitating, isothermal fluid, responds to a fixed gravitational potential of a bar, disc and spheroid, and is modeled with an Eulerian code on a fixed grid. The density is shown in grayscale. Arrows mark gas velocity in the reference frame rotating with the bar. Dotted circles mark corotation and the Inner Linblad Resonance. Examples of x_1 and x_2 orbits are drawn with solid and dashed lines, respectively. "PS" marks the principal shock in the bar; "NR" is the nuclear ring. Units on the axes are in kpc. The insert shows flow pattern of gas with higher velocity dispersion in the same potential. Here density is marked by contours, at 1.4, 2, 3, 5, 7 and 9 times the initial value. The shock indicator, $\mathrm{div}^2\ \hat{\mathrm{v}}$ for $\mathrm{div}\ \hat{\mathrm{v}} < 0$, is shown in grayscale in the insert. From models S05 and S20 by Maciejewski et al. (2002).

(large negative div $\hat{v}$), which is indicative of a shock. Significant gas inflow to the galactic centre may occur if the spiral shock proves to be efficient in transferring angular momentum from the gas to the bar. Recent high-resolution maps of galactic centres often reveal nuclear spirals extending down to the HST resolution limit (e.g. Pogge & Martini 2002).

4. Nested Bars

Nested bars (a secondary bar inside the main one) are quite common phenomenon in galaxies (e.g., Erwin & Sparke 2002). If gas flow in the secondary bar were analogous to the one in the main bar, then the secondary bar would force gas deeper into the potential well. However, these flows are unlikely to be analogous, as Maciejewski & Sparke (2000) showed by exploring limitations imposed on nested stellar bars by their orbital structure. Maciejewski et al. (2002) confirmed this with hydrodynamical simulations: gas flow in the inner stellar bar, which is in resonant coupling with the large bar (see Sygnet et al. 1988), does not develop straight shocks, and does not increase the inflow. It instead organizes itself in an ellipsoidal pattern around the inner bar, like the one seen in color maps of NGC 3081 (Regan & Mulchaey 1999). Nevertheless, if nested bars are not in resonant coupling, other modes of gas flow in the inner bar are possible, including straight dust lanes (Maciejewski 2002).

References

Athanassoula, E. 1992, MNRAS, 259, 345

Benedict, G.F., Howell, D.A., Jørgensen, I., Kenney, J.D.P. & Smith, B.J. 2002, AJ, 123, 1411

Binney, J. & Tremaine, S. 1987, Galactic Dynamics (Princeton Univ. Press: Princeton)

Contopoulos, G. & Papayannopoulos, T. 1980, A&A, 92, 33

Erwin, P. & Sparke, L.S. 2002, AJ, 124, 65

Eskridge, P.B. et al. 2000, AJ, 119, 536

Maciejewski, W. 1998, Ph.D. thesis, University of Wisconsin

Maciejewski, W. 2002, ASP Conf. Ser. 275: Disks of Galaxies: Kinematics, Dynamics and Perturbations, p. 251

Maciejewski, W. & Sparke, L.S. 2000, MNRAS, 313, 745

Maciejewski, W., Teuben, P.J., Sparke, L.S. & Stone, J.M. 2002, MNRAS, 329, 502

Maoz, D., Barth, A.J., Ho, L.C., Sternberg, A. & Filippenko, A.V. 2001, AJ, 121, 3048

Piner, B.G., Stone, J.M. & Teuben, P.J. 1995, ApJ, 449, 508

Pogge, R.W. & Martini, P. 2002, ApJ, 569, 624
Regan, M.W., Vogel, S.N. & Teuben, P.J. 1997, ApJ, 482, L143
Regan, M.W. & Mulchaey, J.S. 1999, AJ, 117, 2676
Sandage, A. 1961, The Hubble Atlas of Galaxies (Carnegie Institution: Washington)
Sørensen, S.-A., Matsuda, T. & Fujimoto, M. 1976, Ap&SS, 43, 491
Sygnet, J.F., Tagger, M., Athanassoula, E. & Pellat, R. 1988, MNRAS, 232, 733

International Symposium on Astrophysics Research and on the Dialogue between Science and Religion, Vatican Observatory, 2003
C. D. Impey & C. E. Petry, editors

Causation Existence and Creation in Space-Times with Non-Trivial Topology

Gustavo E. Romero

Instituto Argentino de Radioastronomía, C.C.5, (1894) Villa Elisa, Buenos Aires, Argentina

Abstract. The Kalam Cosmological Argument is perhaps the most solid and widely discussed argument for a caused creation of the universe. The usual objections to the argument mainly focus on the second premise. In this paper we discuss the dependency of the first premise on the topological structure of the space-time manifold adopted for the underlying cosmological model. It is shown that in chronology-violating space-times the first premise is also violated. The chronology-violation, in turn, requires a massive violation of the so-called energy conditions which could have observational effects that are briefly discussed here. Hence, astronomical observations could be relevant for the validity of the metaphysical argument. In this sense, it is possible to talk of "observational theology."

1. Introduction

The so-called Kalam Cosmological Argument (e.g. Craig 1979) is a version of the classical cosmological argument based on some medieval Islamic arguments against the infinitude of the past. In modern syllogistic form it can be formulated as follows:

1. Whatever begins to exist has a cause of its existence.
2. The Universe began to exist.
3. Therefore, the Universe has a cause of its existence.

It has been argued that the first premise is a fundamental metaphysical principle which cannot be intelligibly denied and that the second premise is supported by modern cosmology, in such a way that the conclusion of the argument is true (Craig 1979, Craig & Smith 1993). These contentions have been discussed in recent years by several philosophers, notably Adolf Günbaum, who argued that the Big Bang model does not support the second premise (e.g. Grünbaum 1989, 1990, 1991, 2000, and replies in Craig 1991 and 1992). The first premise, on the contrary, has not been considered controversial except from the point of view of quantum mechanics (see the discussions in Craig & Smith 1993).

In this paper we shall argue that the validity of the first premise depends on the topology of space-time manifold adopted for the cosmological model. Multiple connected space-times can be compatible with objects that obey all physical laws but violate the first premise of the Kalam Cosmological Argument. Some semantic comments are in order first to clarify the meaning of the expression "to begin to exist."

2. A Semantical Note

Craig (1992) attributes to Grünbaum the implicit use of the following definition: "x begins to exist"$=$ $3D$def. "x exists at time t and there are instants of time immediately prior to t at which x does not exist."

This definition is objected to because it is difficult to accept that the existence of x at t can entail the existence of temporal instants prior to t. Admittedly, in the context of a relational theory of space-time (e.g. Perez-Bergliaffa, Romero & Vucetich 1998), the requirement of the existence of moments prior to t is nothing else than the requirement of the existence of objects other than x *before* x. Such a definition, then, is not adequate to the discussion of the origin of the system formed by *all* things, i.e. the universe. Craig, in turn, proposes: "x begins to exist"$=$ $3D$def. "x exists at time t and there are no instants of time immediately prior to t at which x exists."

This allows for a beginning of time itself and is apparently apt for a discussion of the beginning of the universe. But it also has the problem of demanding a sharp edge for the existence of x. Anything created by an evolutionary process lasting a finite time interval is excluded. Let us consider, for example, humans. They certainly exist now and they certainly did not exist 50 million years ago, but can we point out an instant t at which they did exist and an immediately prior instant at which they did not? Not only biological counter-examples are possible, but we can think in terms of most physical systems, like a star or a molecular cloud, which are formed by a slow transition from a previous state.

In order to remove this problem we propose: "x begins to exist"$=$ $3D$def. "x exists at time t and there is a time interval $\Delta t \geq 0$ such that there are no instants of time immediately prior to $t - \Delta t$ at which x exists."

For $\Delta t = 3D0$, we recover Craig's definition. In what follows we shall understand "to begin to exist" in the sense of this latter definition.

3. Chronology-Violating Space-Times and Self-Existent Objects

A relativistic space-time is represented by a four-dimensional manifold M equipped with a Lorentzian metric g_{ab}. The general theory of relativity requires the manifold to be continuous and differentiable but not specific constraints are imposed on the details of its topology. Usually, simply connected manifolds are considered, but multiply connected ones cannot be ruled out only on *a priori* grounds.

In recent years there has been a sustained interest in multiply connected space-times, also called wormhole space-times, originated in the fact that close timelike curves (CTCs) naturally appear in them (e.g. Morris, Thorne & Yurtsever 1988, Thorne 1992). These curves represent the world lines of any physical system in a temporally orientable space-time that, moving in the future direction, arrives back at some point of its own past. Any space-time with CTCs is called a chronology-violating space-time. Objections to the formation of CTCs in the real universe had been formulated by a number of scientists, most notably by Hawking (1992), but in the absence of a theory of quantum gravity the possibility of wormholes in space-time cannot be ruled out (see discussions and references in Earman 1995a, Romero & Torres 2001, and Nahin 1999).

One of the most strange implications of chronology-violating space-times is the possibility of an ontology with self-existent objects. These are physical systems "trapped" in CTCs. Romero & Torres (2001), who have discussed these systems in depth, give the following toy example to illustrate the nature of such objects:

> *Suppose that, in a space-time where CTCs exist, a time traveler takes a ride on a time machine carrying a book with her. She goes back to the past, forgets the book in what will be her laboratory, and returns to the future. The book remains then hidden until the time traveler finds it just before starting her time trip, carrying the book with her.*

The book in question is a self-existent object: it exists at a given t, there exists $\Delta t \geq 0$ such that the object does not exist at $t - \Delta t$, but, however, there is no external cause of its existence. The self-existent object is *just a feature of space-time itself*, it is not either created or destroyed *in* space-time. Such objects clearly violate the first premise of the Kalam Cosmological Argument.

It is very important to emphasize that, despite the fact that the self-existent objects have not a cause of their existence, they do not violate causality. In fact, since their space-time history is a continuous closed curve, *their physical state at every time t is casually linked to a previous state.* In this way, these objects are not causally created, but they have a finite existence in the sense that they exist during a finite time interval, and their existence does not violate strict causality.

Romero & Torres (2001) have argued against an ontology of self-existent objects invocating a full Principle of Self-Consistency for *all* laws of nature. This principle, which is used to dissolve the so-called "paradoxes" of time travel (Earman 1995b, Nahin 1999), can be stated as:

> *The laws of nature are such that any local solution of their equations that represents a feature of the real universe must be extendible to a global solution.*

Romero and Torres suggest that this principle is a *metanomological* statement (see Bunge 1961) that enforces the harmony between local and global affairs in space-time. By including thermodynamics in the consistency analysis of the motion of macroscopic systems through wormhole space-times, they have shown that *non-interacting* self-existent objects are not possible in the real universe because energy degradation along the CTC results in non-consistent histories.

Notwithstanding these objections, the development of consistent histories remains an open possibility for isolated systems where entropy cannot be defined (e.g. single particles) and for interacting systems where their energy degradation is exactly compensated by external work made upon them (Lossev & Novikov 1992). Hence, if CTCs actually occur in the universe, there seems to be no form to avoid the possibility of at least some types of self-existent objects.

Very recently, Gott and Li (1998) have even proposed that the universe itself could be a self-existent object. From a philosophical point of view, this would be a violation of both premises of the Kalam Cosmological Argument with a single counterexample. As far as it can be seen, the work by Gott and Li is consistent with the Big Bang paradigm. They only require the existence of a multiply connected space-time with a CTC region beyond the original inflationary state.

A key point for the validity of the first premise of the Kalam Cosmological Argument is that the space-time in the real universe must be described by a simply connected manifold, with no CTCs.[1] Otherwise, the presence of objects that have "began to exist" without external cause but notwithstanding are subject to causality cannot be excluded. We have, then, two possibilities in order to explore the validity of the first premise in the context of its dependency on the underlying topology of space-time: (1) we can try to prove, from basic physical laws, that CTCs cannot be formed in the real universe (i.e. we can try to find out a mechanism to enforce chronology protection), or (2) we can inquire about the observational

[1]Formally, CTCs are possible even in simply connected space-times, but these kind of solutions of Einstein field equations, like the classical Gödel (1949) rotating universe, are thought to be not applicable to the real world.

signatures of wormhole structures in space-time and try to test through observations the hypothesis that natural wormholes actually do exist. The first option requires a full theory of quantum gravity, something that is beyond our present knowledge. The second approach is being already explored by some scientists.

4. Observational Signatures of WEC-Violating Matter

Macroscopic and static wormhole structures as those necessary to allow the formation of CTCs require that the average null energy condition (ANEC) be violated in the wormhole throat. This condition is part of the so-called energy conditions of Einstein gravity, which are very general hypotheses designed to provide as much information as possible on a wide variety of physical systems without specifying a particular equation of state. These conditions are not proved from basic principles, they are just conjectures, which can be very useful in some contexts. However, many violating systems are known, including the universe itself (see Visser 1996).

The energy conditions violated by traversable wormhole can be put in terms of the stress-energy tensor of the matter threading the wormhole as $\rho + p \geq 0$, where ρ is the energy density and p is the total pressure. This implies also a violation of the so-called weak energy condition (WEC or $\rho \geq 0 \ \wedge \ \rho + p \geq 0$; see Visser 1996 for details, also Morris and Thorne 1988). Plainly stated, all this means that the matter threading the wormhole must exert gravitational repulsion in order to stay stable against collapse. If natural wormholes exist in the universe (e.g. if the original topology after the Big Bang was multiply connected), then there should be observable signatures of the interactions between matter with negative energy density with the normal matter.

At astronomical levels, the most important observational consequence of the existence of natural wormholes is gravitational lensing of background sources (Cramer et al. 1995, Torres et al. 1998; Eiroa et al. 2001, Safonova et al. 2001). There are very specific features produced by chromaticity effects in lensing of extended sources that could be used to differentiate events produced by wormholes from those of other objects (Eiroa et al. 2001). In the wormhole microlensing case there are two intensity peaks during each event separated by an umbra region. On the contrary, in the normal case there is a single, time-symmetric peak. In addition, in the wormhole case it can be shown that there is a spectral break that is not observed in the usual case (see Eiroa et al. 2001 for details).

Also, the macrolensing effects upon a background field of galaxies produced by large-scale violations of the energy conditions are observationally distinguishable from the normal macrolensing by either dark or luminous matter concentrations (see Safonova et al. 2001 for complete numerical simulations of macrolensed galaxy fields). In particular, it can be shown that for positive mass we see concentric arcs, whereas for negative energy densities we have filamentary features projected from the center.

The above examples are enough to illustrate the kind of observational effects that can be expected in an universe with multiple connected topology. Whether such space-time wormholes actually exist in our universe is something that has to be found yet.

The mere existence of a multiple connected topology for space-time does not warrant, by itself, the violation of the first premise of the Kalam Cosmological Argument. But it makes possible the formation of CTCs and non-chronal situations in that space-time, hence opening the possibility of an ontology with self-existent objects. This implies that the universality of the premise can be objected even at a macroscopic level, without resorting to quantum considerations.

5. Conclusions: Theology meets experiment

The first premise of the Kalam Cosmological Argument, namely that "whatever that begins to exist has a cause of its existence," is not a self-evident, universally valid statement as it is usually accepted. We have shown that the truth value of the premise is dependent on some basic characteristics of the space-time manifold that represents the real universe. In particular, multiple connected space-times can accommodate objects that exist by themselves, without external cause, but also without any local violation of causality. These objects "begin to exist" in accordance to even the most restrictive definitions given in Section 2.

Since the connectivity of space-time can be probed through astronomical observations (see Anchordoqui et al. 1999 for an example of these observational studies), the validity of the Kalam Cosmological Argument can be tested by the scientific method. Not only the second premise, which uses to be discussed in the light of the Big Bang cosmology, but also the first premise of the argument is susceptible to experimental test. It is in this more extended sense that in the Kalam Cosmological Argument we can say that theology meets experiment.

Acknowledgments. I thank Diego F. Torres and Santiago E. Perez-Bergliaffa for many illuminating discussions on space, time, and causality along many years. This work has been supported by the agencies CONICET and ANPCT (PICT 03-04881) as well as by funds granted by Fundación Antorchas.

References

Anchordoqui, L.A., Romero, G.E., Torres, D.F. & Andruchow, I. 1999, Mod. Phys. Lett. A, 14, 791

Bunge, M. 1961, Am. J. Phys., 29, 518

Craig, W.L. 1979, The Kalam Cosmological Argument (Harper & Row: New York)

Craig, W.L. 1991, Nature, 354, 347

Craig, W.L. 1992, Brit. J. Phil. Sci., 43, 233

Craig, W.L. & Smith, Q. 1993, Theism, Atheism, and Big Bang Cosmology (Clarendon Press: Oxford)

Cramer, J.G. et al. 1995, Phys. Rev. D, 51, 3117

Earman, J. 1995a, Erkenntnis, 42, 125

Earman, J. 1995b, Bangs, Crunches, Whimpers, and Shrieks (Oxford University Press: New York)

Eiroa, E., Romero, G.E. & Torres, D.F. 2001, Mod. Phys. Lett. A, 16, 973

Gödel, K. 1949, Rev. Mod. Phys., 21, 447

Gott, J.R. & Li, L.X. 1998, Phys. Rev D, 58, 235

Grünbaum, A. 1989, Phil. Sci., 56, 373

Grünbaum, A. 1990, Nature, 344, 821

Grünbaum, A. 1991, Erkenntnis, 35, 233

Grünbaum, A. 2000, Brit. J. Phil. Sci., 51, 1

Hawking, S.W. 1992, Phys. Rev. D, 46, 603

Lossev, A. & Novikov, I.D. 1992, Class. Quantum. Grav. 9, 2309

Morris, M.S. & Thorne, K.S. 1988, Am. J. Phys., 56, 395

Morris, M.S., Thorne, K.S. & Yurtsever, U. 1988, Phys. Rev. Lett., 61, 1446.

Nahin, P.J. 1999, Time Machines: Time Travel in Physics, Metaphysics and Science Fiction (Springer-Verlag and AIP Press: New York)

Perez-Bergliaffa, S.E., Romero, G.E. & Vucetich, H. 1998, Int. J. Theoret. Phys., 37, 2281

Safonova, M., Torres, D.F. & Romero, G.E. 2001, Mod. Phys. Lett. A, 16, 153

Thorne, K.S. 1992, in General Relativity and Gravitation 1992, eds. J.L. Gleiser et al. (Institute of Physics: Bristol)

Torres, D.F., Romero, G.E. & Anchordoqui, L.A. 1998, Phys. Rev. D, 58, 1231

Visser, M. 1996, Lorentzian Wormholes (AIP Press: New York)

International Symposium on Astrophysics Research and on the Dialogue between Science and Religion, Vatican Observatory, 2003
C. D. Impey & C. E. Petry, editors

An Infrared View of Blue Dwarf Galaxies

Leonardo Vanzi

European Southern Observatory, Alonso de Cordova 3107, Vitacura, Santiago de Chile

Abstract. In this paper, I review the infrared properties of blue dwarf galaxies. Recent observations of blue dwarf galaxies in the infrared have generated extremely interesting results, opening an entirely new field of research. Near-infrared observations have detected evolved stellar populations in most blue dwarfs though a few objects still hold promise as possible genuine young galaxies. Observations in the mid-infrared and millimetre regimes have revealed in a few cases unexpected amounts of dust and the presence of embedded young stellar clusters. This findings challange our understanding of the star formation and shed new light on the early evolution of galaxies.

1. Introduction

Blue dwarf galaxies are low-mass extragalactic objects where intense star formation is presently occurring, as evidenced by their blue UBV colors and their optical spectra. Ever since their discovery they have been considered excellent candidates for being young galaxies, i.e., systems presently undergoing one of their very first bursts of star formation (Searle & Sargent 1972). This is mainly because blue dwarf galaxies are very metal-deficient with metallicities ranging from $Z_{\odot}/3$ to $Z_{\odot}/50$. Other important evidence of their relative youth is the very high fractional neutral hydrogen gas content (e.g., Thuan & Martin 1981, Pustilnik et al. 2001) and the lack of an evident underlying old stellar population in optical images. Blue Dwarfs therefore represent an extremely valuable tool for studying details of the process of star formation in a primordial environment, a critical topic for our understanding of the early Universe.

2. An Infrared View

Observations in the near infrared can help to solve the age problem. The bands around 1 μm are sensitive to the presence of evolved stars. The first near-infrared observations of blue dwarf galaxies (Thuan 1983, 1985) proved that in the majority of the cases an underlying evolved stellar pop-

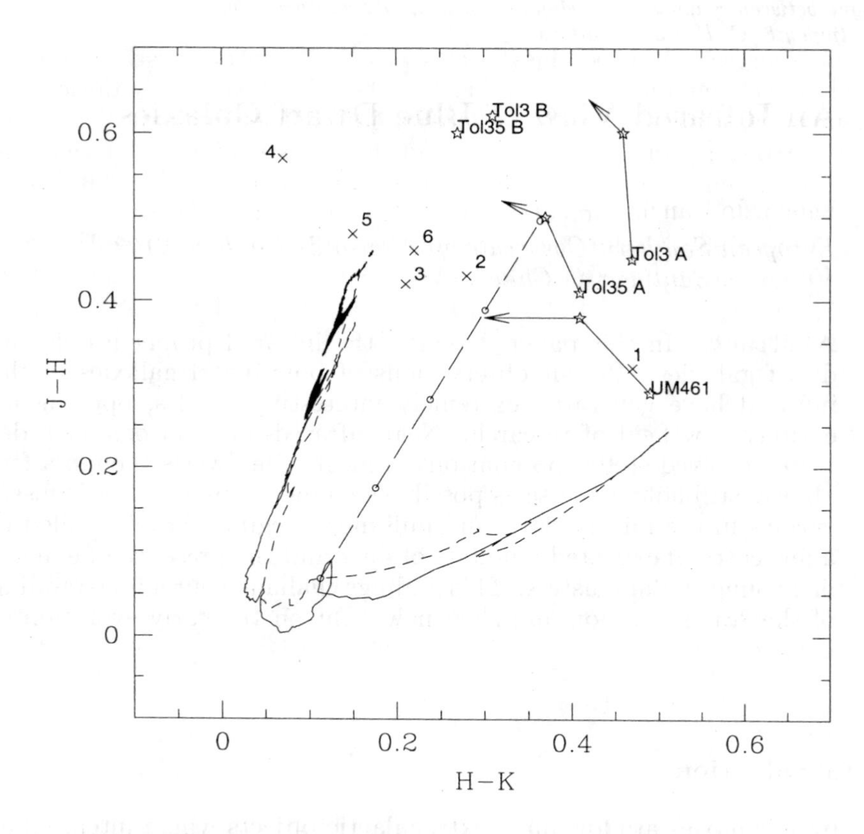

Figure 1. Near-infrared colors of three blue dwarf galaxies. The arrows indicate the correction applied to the observed values to take into account the effect of the emission lines and gas continuum. The model SB99 is plotted as solid and dashed line. The long dashed line indicates the effect of the extinction, with marks separated by $A_V = 1$.

ulation is present. Most Blue Dwarfs are therefore old systems. A few cases however remain quite puzzling. In particular, recent deep optical and near-infrared studies of the three most metal-deficient galaxies—I Zw 18, SBS 0335–052 E, and W—gave no evidence for stellar populations older than a few hundred Myr (Papaderos et al. 2002, Vanzi et al. 2000, Pustilnik et al. 2002). The indications for a 1 Gyr old population in I Zw 18 (Aloisi et al. 1999, Ostlin 2000) are not widely accepted because the interpretation of near-infrared and optical colors of very young stellar systems is heavily affected by the radiation emitted by the gas (Vanzi et al. 2002). Spectroscopic observations can help to correct for this effect. This is shown in Figure 1, where the observed and gas-corrected near-infrared colors of a few blue dwarfs are shown.

Among the first observations obtained at longer wavelengths are the 10 μm spectra of He 2-10, II Zw 40 and NGC 5253 of Roche et al. (1991). They found evidence for HII regions and high excitation starburst features. A real surprise however came with the ISO observation of SBS 0335-052 by Thuan et al. (1999); their mid-infrared spectrum is basically featureless and dominated by a very bright continuum. This was interpreted as indicating the presence of a large amount of dust—totally unexpected at such a low metallicity ($Z_{\odot}/40$). Hunt et al. (2001) detected a very bright emission from the Brα recombination line which, combined with Brγ, gives a visual extinction exceeding 12 mag. The star formation rate derived from this observation at 4μm is 75% higher than what is derived from optical data. Plante & Sauvage (2002) constrained the spectral energy distribution of SBS 0335-052 from the near-infrared to the millimetre, finding an extinction of about 30 magnitudes in V, and a large amount of dust—about 10^5 $M_{\odot}$. Such dust content is quite puzzling in a low metallicity environment as a metal-poor medium is expected to be relatively dust free. Hirashita et al. (2002) attempt to attribute this dust to the SNe generated by the current burst of star formation. SNe could indeed generate enough dust to explain the observations, however the processes involved in the dust balance are not very well understood.

Other interesting cases have been appearing in the very last years. He 2–10, for instance, was first imaged in the mid-infrared by Sauvage et al. (1997), revealing emission by dust. In the same galaxy, Vacca et al. (2002) observed at 10μm young clusters completely obscured in the optical but detected in the radio (Kobulnicky & Johnson 1999). The radio and infrared morphologies were consistent. Similarly embedded clusters have been observed in the blue dwarf galaxy NGC 5253 in the radio (Turner et al. 2000) and infrared-millimetre (Vanzi & Sauvage 2003). From the modeling of the spectral energy distribution of these objects over a large spectral range, many interesting properties can be derived including extinction, dust mass, size distribution and composition, age, etc. The properties of the dust observed so far seem to be non-standard.

3. Conclusions

Observations in the near-infrared first succeeded in detecting an evolved stellar population in most blue dwarf galaxies, however, a handful galaxies still elude the attempts in this direction. Some blue dwarf galaxies are revealing surprising properties when observed in the infrared. They show unexpected amount of dust and in some cases an interesting class of dust involved with embedded massive young clusters.

References

Aloisi, A., Tosi, M. & Greggio, L. 1999, AJ, 118, 302
Hirashita, H., Hunt, L.K. & Ferrara A. 2002, MNRAS, 330, L19
Hunt, L.K., Vanzi, L. & Thuan, T.X. 2001, A&A, 377, 66
Kobulnicky, H.A. & Johnson, K.E. 1999, ApJ, 527, 154
Ostlin, G. 2000, ApJ, 535, L99
Papaderos, P., Izotov, Y.I., Thuan, T.X. et al. 2002, A&A, 393, 461
Plante, S. & Sauvage, M. 2002, AJ, 124, 1995
Pustilnik, S.A. et al. 2001, AJ, 121, 1413
Pustilnik S.A. et al. 2002, AJ, submitted
Roche, P.F., Aitken, D.K. & Smith, C.H. 1991, MNRAS, 252, 282
Sauvage, M., Thuan, T.X. & Lagage, P.O. 1997, A&A, 325, 98
Searle, L. & Sargent, W.L.W. 1972, ApJ, 173, 25
Thuan, T.X. & Martin G.E. 1981, ApJ, 247, 823
Thuan, T.X. 1983, ApJ, 268, 667
Thuan, T.X. 1985, ApJ, 299, 881
Thuan, T.X., Sauvage, M. & Madden, S. 1999, ApJ516, 783
Turner, J.L., Beck, S.C. & Ho, P.T.P. 2000, ApJ, 532, L109
Vacca, W.D., Johnson, K.E. & Conti, P.S. 2002, AJ, 123, 772
Vanzi, L., Hunt, L.K., Thuan, T.X. & Izotov, Y.I. 2000, A&A, 363, 493
Vanzi, L., Hunt, L.K. & Thuan, T.X. 2002, A&A 390, 481
Vanzi, L. & Sauvage, M. 2003, A&A, submitted

International Symposium on Astrophysics Research and on the Dialogue between Science and Religion, Vatican Observatory, 2003
C. D. Impey & C. E. Petry, editors

Searching for the Debris of Ancient Galaxies in the Milky Way's Halo

Anna Katherina Vivas

Centro de Investigaciones de Astronomía (CIDA), Apartado Postal 264, Mérida 5101-A, Venezuela

Abstract. The formation and evolution of galaxies is one of the major unsolved problems in astrophysics. Clues to this problem may be found by examining our galaxy, particularly the "fossil" record of the early stages of galactic evolution preserved in the galactic halo stellar population. If the halo has been built by accretion events, we do not expect it to have a smooth distribution of stars. The halo may consist of streams of stars that originated in now extinct satellite galaxies. Our survey of RR Lyrae stars is aimed to find such streams in the outer halo of the Milky Way. Several sub-structures have been detected including one which seems to be tidal debris from the Sagittarius dwarf spheroidal galaxy.

1. Introduction

Hierarchical models of galaxy formation predict that galaxies were formed by the merger of small units, such as dwarf spheroidal galaxies. Such mergers would disrupt the small galaxies, leaving a trail of debris along their orbital paths. Numerical simulations have shown that the streams of debris have long lives, especially in the outer halo (Johnston 1998, Helmi & White 1999). Thus, the halo of the Milky Way may consist of streams of stars that originated in now extinct satellite galaxies. The detection and quantification of streams provide an indicator of how important the role of accretion was in the formation of the halo.

The search for substructures in the halo has been difficult in the past because it is necessary to survey a large area of the sky (hundreds of square degrees) to faint limiting magnitudes. The advent of large format CCD cameras has made this possible and several clumps of stars in the halo have been detected (Yanny et al. 2000, Vivas et al. 2001, Newberg et al. 2002, Vivas 2002).

We present here the results of a large scale survey for RR Lyrae stars aimed at studying the structure of the halo. Those stars are good tracers of the old stellar population of the halo. The survey is designed to detect

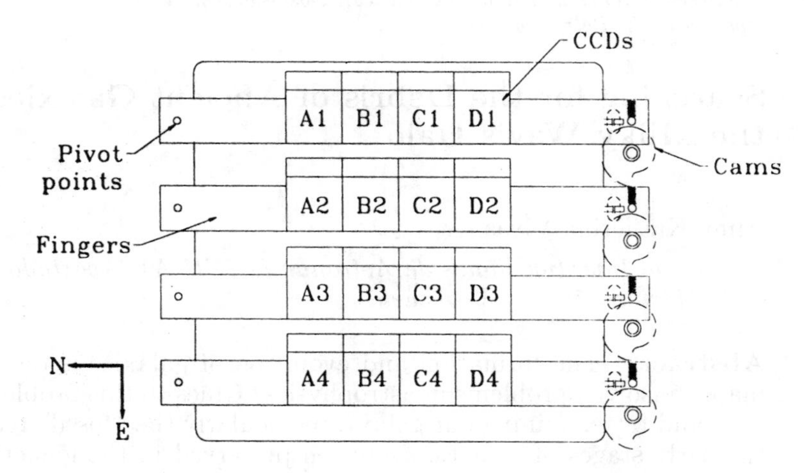

Figure 1. Diagram of the 16-chip QUEST camera.

numerous RR Lyrae stars at large distances from the Galactic Center, a region which has been poorly studied in the past.

2. The Survey for RR Lyrae Stars

The survey was carried out with the QUEST Camera at the 1m Schmidt telescope in the Llano del Hato Observatory, Venezuela. The camera is a mosaic of sixteen 2048 × 2048 CCDs (Figure 1), with a field of view of $2\overset{\circ}{.}3 \times 2\overset{\circ}{.}3$. It was built by the QUEST (Quasar Equatorial Survey Team) Collaboration, which includes a large number of scientists at institutions in Venezuela and the United States.

The camera has been designed for observing in drift-scan mode: the telescope is fixed and the CCDs are read out in the East-West direction, at the sidereal rate as stars drift across the device. This procedure results in the generation of a continuous strip of the sky, $2\overset{\circ}{.}3$ wide. The drift-scan mode allows us to obtain quasi-simultaneous multi-band photometry since each row of CCDs can be fitted with separate filters; stars will successively cross over 4 chips in a column each with its own filter. The system is very efficient for covering large areas in a short time: one can survey the sky at a rate of 34.5 square degrees per hour per filter. More technical details on the camera can be found in Baltay et al. (2002).

For the first part of the survey (Vivas 2002), we were able to combine more than 100 observations from different projects. Each of these projects required the repeated observation of the same large region of the sky to the faintest possible limiting magnitude. We covered a total of 380 deg^2 in a strip centered at $\delta = -1°$. In Right Ascension, we covered the regions $4^{\rm h} < \alpha < 6^{\rm h}$ and $8^{\rm h} < \alpha < 17^{\rm h}$. Thus, we span a large range in galactic

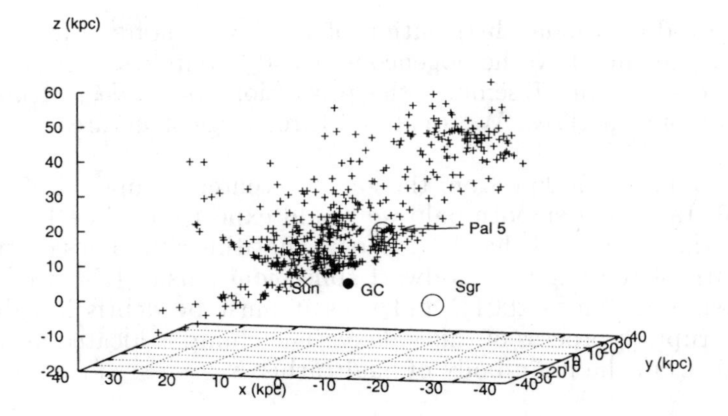

Figure 2. Three-dimensional view of the location of the RR Lyrae stars (+). The position of the Sun, the Galactic center, the globular cluster Pal 5 and the Sagittarius dwarf spheroidal galaxy are indicated.

coordinates, from $|b| = 10°$ to $63°$. There are between 15 and 40 observations along each different line of sight. The limiting magnitude, $V \sim 19.7$, depends on the exposure time which is set by the time a star needs to cross a single chip (~ 140 seconds). Due to the very high amount of data (~ 1.7 Terabytes), all the basic reduction, astrometry and aperture photometry was done with a custom made automatic pipeline.

The large number of epochs along each line of sight allowed us to search for RR Lyrae stars by calculating the periods of variable objects in the survey and inspecting their phased light curves. Previous constraints in color and maximum amplitude helped to clean up the candidate list. We found a total of 497 RR Lyrae stars, 86% of which are newly discovered objects.

3. Sub-Structures in the Halo

RR Lyrae stars are well known standard candles. We calculated distances to the stars in our survey in order to study their spatial distribution in the halo. The combination of large area and depth of the survey provides new insights on the structure of the halo, especially in the outermost parts. Before this survey, very few objects were known at distances larger than 30 kpc from the galactic center.

The 3-dimensional distribution of RR Lyrae stars (Figure 2) shows the halo does not have homogeneous density contours. The distribution is quite clumpy and resembles the predictions of models of hierarchical formation of structures. We identified three large structures:

1. A clump at 50 kpc from the galactic center, composed of more than 80 stars (in a smooth halo we would expect only 14 RR Lyrae stars in that region). It has been shown that this clump lies on the orbital path of the Sagittarius dwarf spheroidal galaxy (Ibata et al. 2001, Helmi & White 2001), and thus it must be debris left during the disruption process of this galaxy. The clump is located at more than 60° from the main body of Sagittarius.

2. There is an over-density of RR Lyrae stars in the neighborhood of the globular cluster Palomar 5, at 22 kpc from the Sun. Since this group of stars is well outside the tidal radius of the cluster, it is probably evidence of the disruption of Pal 5 by the tidal forces of the Milky Way.

3. The last group has uncertain origin since it does not seem to be associated with any known dwarf galaxy or globular cluster. It is located at 19 kpc from the Sun. The radial velocities and metallicities of these stars are being measured, and they should shed more light on the origin of this clump.

Acknowledgments. This research was done in close collaboration with Robert Zinn (Yale University, USA). The Llano del Hato Observatory is operated by CIDA for the Ministerio de Ciencia y Tecnología of Venezuela. This research was partially supported by the National Science Foundation under grant AST-0098428. We thank all members of the QUEST collaboration for their important contribution to this work.

References

Baltay, C. et al. 2002, PASP, 114, 780

Helmi, A. & White, S.D.M. 1999, MNRAS, 307, 495

Helmi, A. & White, S.D.M. 2001, MNRAS, 323, 529

Ibata, R. et al. 2001, ApJ, 551, 294

Johnston, K.V. 1998, ApJ, 495, 297

Newberg, H.J. et al. 2002, ApJ, 569, 245

Vivas, A.K. et al. 2001, ApJ, 554, L33

Vivas, A.K. 2002, PhD thesis, Yale University

Yanny, B. et al. 2000, ApJ, 540, 825

International Symposium on Astrophysics Research and on the Dialogue between Science and Religion, Vatican Observatory, 2003
C. D. Impey & C. E. Petry, editors

An X-Ray View of Cosmic Star Formation History

Y.P. Wang

Purple Mountain Observatory, Academia Sinica, China; National Astronomical Observatories of China

T. Yamada

National Astronomical Observatory of Japan, Mitaka, Japan

Y. Taniguchi

Tohoku University, Aramaki, Aoba, Sendai 980-8578, Japan

Abstract. Based on a joint evolutionary scheme of the black hole (BH) growth and the bulge formation, we use the X-ray luminosity function (XLF) of AGNs and their cosmological evolution from ROSAT X-ray surveys to estimate the cosmic star formation history associated with accretion onto the central black holes.

1. Introduction

A mean value of the black hole to bulge mass ratio $M_{bh}/M_{bulge} \sim 10^{-3}$ and the local BH density $\rho_{bh} \sim 3 - 5 \times 10^5$ $M_\odot$ Mpc^{-3} are given by various observations (Salucci et al. 1999, Kormendy & Gebhardt 2001, Merritt & Ferrarese 2001). This correlation implies a possible scenario where the accretion history in the inner few parsecs and the $\sim$ kpc sized star formation during spheroid formation are closely connected (Wang & Biermann 1998, Wang et al. 2000, Page et al. 2001). Since deep X-ray surveys provide a direct probe of the AGN accretion history, we trace the black hole (BH) growth and the joint star formation with soft X-ray 0.5 to 2 keV surveys (Miyaji et al. 2000).

We adopt in the calculation: (1) a derivation of the BH mass from the AGN X-ray emission by assuming an Eddington ratio ϵ, (2) the black hole to bulge mass correlation, (3) a similar time scale for the BH growth and the intensive star formation populating finally the spheroids based on our previous work. The present day BH density and the SCUBA number counts are used as two important model constraints in the calculation, with the set of cosmological parameters $(\Omega_m, \Omega_\Lambda) = (0.3, 0.7)$ and $H_0 =$ 50 km s^{-1} Mpc^{-1}.

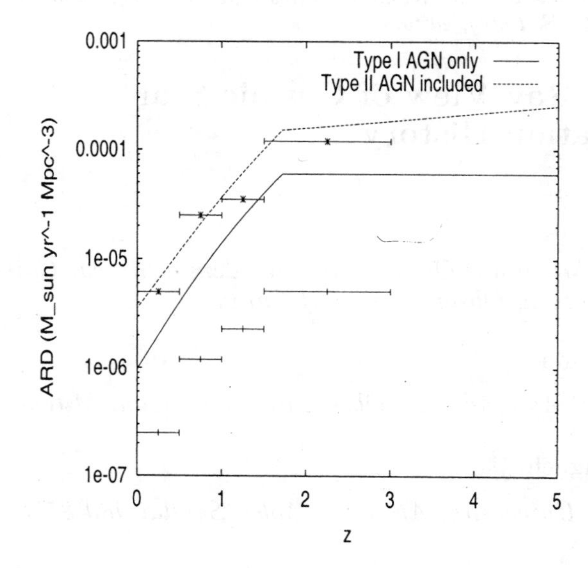

Figure 1. The ARD vs. redshift from the calculation. The blue asterisks represent the ARD calculated from the bolometric luminosities of 69 hard X-ray selected sources, while the red crosses from their X-ray luminosities as a low limit (Barger et al. 2001).

2. BH Growth and Star Formation Viewed with X-Rays

We estimate the BH mass from the X-ray luminosity by assuming an Eddington ratio $\epsilon = \frac{\eta \dot{m} c^2}{L_{\rm Edd}}$. With the bolometric correction β ($L_{\rm bol} = \beta\, L_{\rm x}$), we could get the BH mass $M_{\rm bh} = \frac{\beta\, L_{\rm x}}{0.013\, \epsilon}$. $L_{\rm x}$ is the AGN 0.5 – 2 keV luminosity in units of 10^{40} erg s^{-1} and $M_{\rm bh}$ in units of M$_\odot$. The BH mass function of different redshift is converted from the XLF by assuming reasonable values for the Eddington ratio ϵ and the "duty cycle, $f_{\rm on}$" of the AGN active phase. We adopt $\epsilon = 10^{\gamma\,(\log L\,-\,49)}$; $\gamma \approx 0.2$ is a scaling factor (Salucci et al. 1999). We assume that each AGN shines for a constant time scale $t_{\rm Q} = 5 \times 10^8$ yrs, close to their e-folding time. The "duty cycle" is defined as $f_{\rm on} = t_{\rm Q}/t_{\rm Hub}(z)$, where $t_{\rm Hub}(z)$ is the Hubble time.

We consider all AGNs in Miyaji et al. (2000) samle as type 1 since the fraction of the obscured objects is negligible. $\beta = 20$ is adopted in our calculation based on the mean type 1 AGN spectral energy distribution from Elvis et al. (1994). The BH mass function could be derived from the observed XLF by:

$$\frac{d\Phi(z, M_{\rm bh})}{dM_{\rm bh}} = \frac{0.013\, \epsilon}{\beta\, f_{\rm on}} \frac{\Phi(z, L_{\rm x})}{L_{\rm x}} \tag{1}$$

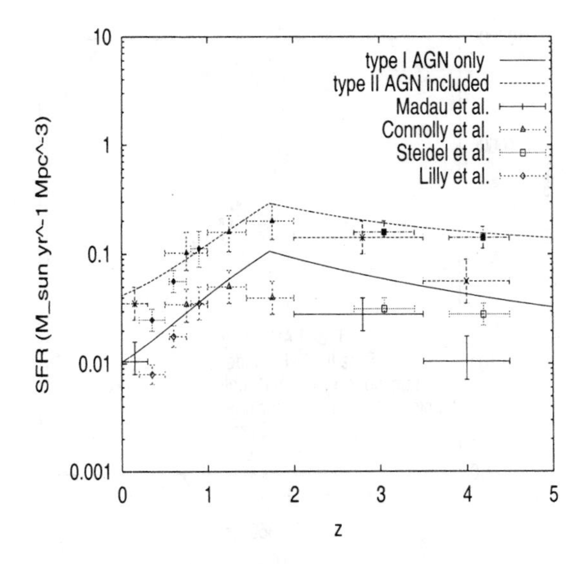

Figure 2. Co-moving SFR in the host galaxies of different AGN sample by the model (with lines). The data sources are indicated in the figure, while the filled symbols are those with dust correction (×4.7) adopted by Steidel et al. (1999).

We follow the LDDE cosmological evolution suggested by Miyaji et al. (2000) for the ROSAT soft XLF, and trace the BH mass density accreted during the active phase with lookback time.

Considering the various studies of the obscured AGNs, we describe the abundance ratio of type 2 to type 1 AGNs including low luminosity Seyferts and QSOs as $R_{2-1} = 4\,e^{-\frac{L_x}{L_s}} + \alpha\,(1+z)^p\,(1 - e^{-\frac{L_x}{L_s}})$. $L_s = 10^{44.3}$ erg s^{-1} is the e-folding luminosity; α and p are free parameters in the model. With the constraints of the present day BH density and the submillimeter deep surveys, we found the ratio of type 2 to type 1 QSOs ($L_x > L_s$) has an upper limit of about two, consistent with the results of recent Chandra deep surveys by Alexander et al. (2001). The accretion rate density (ARD) from the model is shown is Figure 1. It seems that the predicted ARD of type 1 AGNs only is far below the upper bound derived from a multiwavelength study of 69 hard X-ray selected sources in the field of A370, SSA13 and CDF-N (Barger et al. 2001). Meanwhile, the number of type 2 AGNs included in our model agrees with the current deep hard X-ray surveys based on a rough estimation.

Concerning the joint evolution scheme of the massive BH and their spheroids, the mass function of spheroids may have a similar form to the BH mass function and is given by:

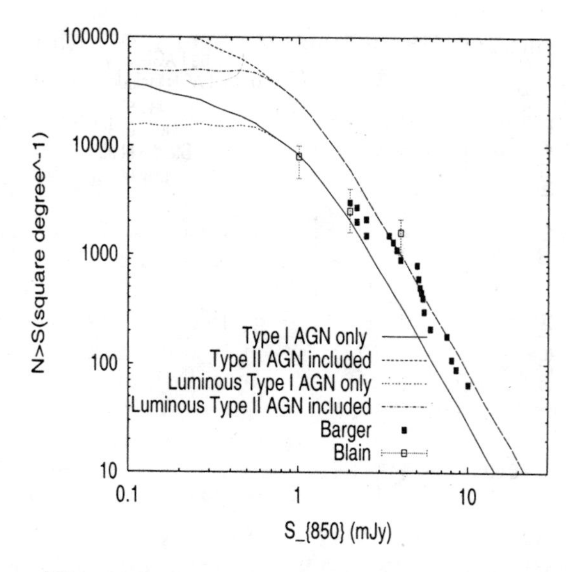

Figure 3. The 850 μm count fitting from the calculation. The data are from Barger et al. (1999) and Blain et al. (1999). We see from the figure that the star formation powered far-infrared emission in the X-ray luminous AGNs ($L > 10^{44}\,\mathrm{erg\,s^{-1}}$) would dominate the SCUBA counts (with dotted lines).

$$\frac{d\Phi(z, M_{\rm sph})}{dM_{\rm sph}} = \frac{d\Phi(z, M_{\rm bh})}{dM_{\rm bh}} \times R, \tag{2}$$

where R is the black hole to bulge mass ratio. We adopt here a mean value $R \sim 0.002$ as the first approximation (Merritt & Ferrarese 2001).

From our previous work, the star formation time scale $\tau_{\rm sf}$ of the starburst activities during BH growth is approximately equal to the QSO life time $t_{\rm Q}$ (i.e., $\tau_{\rm sf} \sim t_{\rm Q} \sim 5 \times 10^8$ yrs). The star formation rate (SFR) in such a phase would be SFR $= \mathrm{M_{sph}}/\tau_{\rm sf}$. The co-moving SFR vs. redshift from our calculation is shown in Figure 2.

To discuss the contribution of the starburst activities associated with BH growth to the SCUBA number counts, we adopt a mean value $L_{\rm FIR}/\mathrm{L_\odot} = 3.8 \times 10^9$ SFR ($\mathrm{M_\odot\,yr^{-1}}$), and a mean color ratio $R_{\rm c} = L_{\rm FIR}/L_{850} \sim 5 \times 10^3$ by Chary & Elbaz (2001) from IRAS, ISO and SCUBA surveys.

Similar to Equation 1, we could derive the 850 μm luminosity function at different redshifts as follows:

$$\frac{d\Phi(z, L_{850})}{dL_{850}} = f_{\widehat{\rm on}} \frac{d\Phi(z, M_{\rm sph})}{dM_{\rm sph}} \frac{dM_{\rm sph}}{dL_{850}}, \tag{3}$$

where $f_{\widehat{\rm on}} = t_{\rm sf}/t_{\rm Hub}$ reflects the fraction of galaxies with intensive star formation ongoing. $\frac{dM_{\rm sph}}{dL_{850}} = \frac{\nu R_{\rm c} t_{\rm sf}}{3.8\times10^{9}{\rm L}_{\odot}}$, $R_{\rm c}$ and $t_{\rm sf}$ are the parameters discussed above. We note that $\nu = 3.5 \times 10^{11}$ Hz is the frequency of 850 μm. The 850 μm number count fitting is shown in Figure 3.

3. Summary

We have shown a possibility to probe the cosmic star formation history associated with AGN accretion by X-ray deep surveys based on the tight correlation of $M_{\rm bh} - M_{\rm B,bulge}$ in early type galaxies and nearby AGNs.

We found, (1) the total amount of star formation associated with the massive BH growth is almost the same as that of Lyman Break Galaxies detected by the current optical deep surveys, although the space density of the especially luminous QSOs is very low; (2) the peak redshift of the spheroid formation is not necessary much beyond 2, evenif we include a reasonable fraction of type 2 QSOs at high redshift; (3) the abundance ratio of type 2 to type 1 QSOs has an upper limit of about 2 from the rough estimation; (4) the far infrared emission from the dust heated by star formation on-going during the BH growth could sufficiently account for the observed SCUBA number counts, and would be the probable dominating energy source of the SCUBA population.

References

Alexander, D.M., Brandt, W.N. et al. 2001, ApJ, 122, 2156

Barger, A.J., Cowie, L.L., Smail, I. et al. 1999, AJ, 117, 2656

Barger, A.J., Cowie, L.L., Bautz, M. W. et al. 2001, AJ, 122, 2177

Blain, A.W., Kneib, J.-P., Ivison, R.J. & Smail, I. 1999, ApJ, 512, L87

Chary, R. & Elbaz, D. 2001, ApJ, 556, 562

Elvis, M. et al. 1994, ApJS, 95, 1

Kormendy, J. & Gebhardt, K. 2001, in XX Texas Symposium on Relativistic Astrophysics, eds., Martel & Weeler (AIP: College Park, MD), in press (astro-ph/0105230)

Merritt, D. & Ferrarese, L. 2001, MNRAS, 320, L30

Miyaji, T., Hasinger, G. & Schmidt, M. 2000, A&A, 353, 25

Page, M.J., Stevens, J.A. et al. 2001, Science, 294, 2516

Salucci, P., Szuszkiewicz, E., Monaco, P. et al. 1999, MNRAS, 307, 637

Steidel, C.C., Adelberger, K.L., Giavalisco, M. et al. 1999, ApJ, 519, 1

Wang, Y.P. & Biermann, P.L. 1998, A&A, 334, 87 (WB98)

Wang, Y.P. & Biermann, P.L. & Wandel, A. 2000, A&A, 361, 550

Part 2. Contributed Papers

Section C. Religion and Smörgasbord

International Symposium on Astrophysics Research and on the Dialogue between Science and Religion, Vatican Observatory, 2003
C. D. Impey & C. E. Petry, editors

Massive Variability Surveys from Venezuela

César Briceño

Centro de Investigaciones de Astronomía (CIDA), Apartado Postal 264, Mérida 5101–A, Venezuela

Abstract. At the Venezuela National Astronomical Observatory we are carrying out variability surveys spanning many hundreds of square degrees near the celestial equator, using an 8k x 8k CCD Mosaic Camera optimized for drift-scanning on a 1m Schmidt telescope. Among the initial efforts was a project to obtain the first moderately deep, homogeneous sample of young stars over an area of $\sim$ 180 square degrees encompassing the entire Orion OB1 association—one of the nearest and most active regions of star formation. The results show that variability is a powerful technique to identify pre-main sequence populations, especially in sparse areas devoid of gas and dust. We are currently developing a massive database, equipped with web-based data mining tools, that will make our data and results available to the astronomical community.

1. Introduction

Recent surveys focusing on repeated observations of large expanses of the night sky (e.g. ROTSE, Akerlof et al. 2000) are transforming the study of variability phenomena into one of the forefront research areas in modern day astrophysics. These searches have also demonstrated the unique value of small telescopes for this kind of research; in particular, the wide field Schmidt design, when equipped with mosaic CCD detectors, is a potentially ideal combination to collect multi-epoch data over extended areas of the sky in an efficient way. In this contribution, I describe one of the variability surveys, spanning many hundreds of square degrees near the celestial equator, that are being carried at the Venezuela National Astronomical Observatory using an 8k x 8k CCD Mosaic Camera, optimized for drift-scanning, installed on a 1m Schmidt telescope.

2. A Variability Survey of the Orion OB1 Association

The study of how stars and planetary systems form and evolve is an area in which variability surveys are particularly useful as a means to search for pre-main sequence objects. Young (1–10 Myr) low mass ($\lesssim 2\ M_{\odot}$)

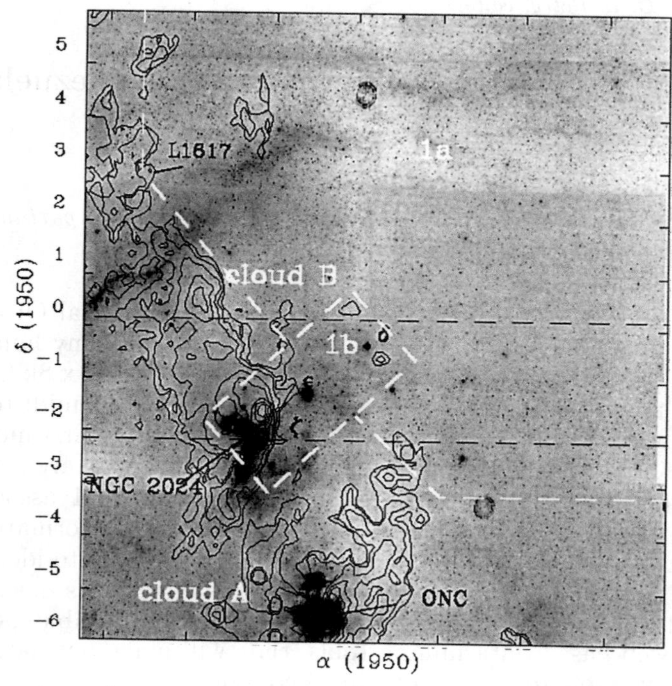

Figure 1. A composite Digital Sky Survey image of Orion showing the total survey area of 180 square degrees. The first strip we observed, covering the three Orion belt stars, is indicated by dashed lines. The Orion Nebula Cluster (ONC) and the nebulosity around the young region NGC 2024 are indicated.

stars are intrinsically variable at both optical (Herbst 1986) and near IR wavelengths (Carpenter, Hillenbrand & Strutskie 2001) by up to several magnitudes, thus variability should help to single out these objects among the general field population. However, variability has been used mainly in follow-up studies rather than as a survey technique so far, to collect time-series data of young objects that were detected by some other method, like color-magnitude diagrams, X-ray and Hα surveys.

In collaboration with Nuria Calvet (CfA), A.K. Vivas (CIDA) and Lee Hartmann (CfA), we have used the 8k $\times$ 8k CCD Mosaic Camera on the Venezuela 1m Schmidt telescope (see Baltay et al. 2002 for a description of the camera; also Vivas in this volume) to conduct a large scale, systematic optical VRIHα survey spanning $\sim$ 180 square degrees in the Orion OB1 Association (Figure 1), one of the nearest ($d \sim 400$ pc, Brown et al. 1994) star forming regions. Though Orion has been extensively studied in the past, existing optical/IR studies in Orion have concentrated mostly on small regions such as the Orion Nebula Cluster (ONC; e.g., Hillenbrand

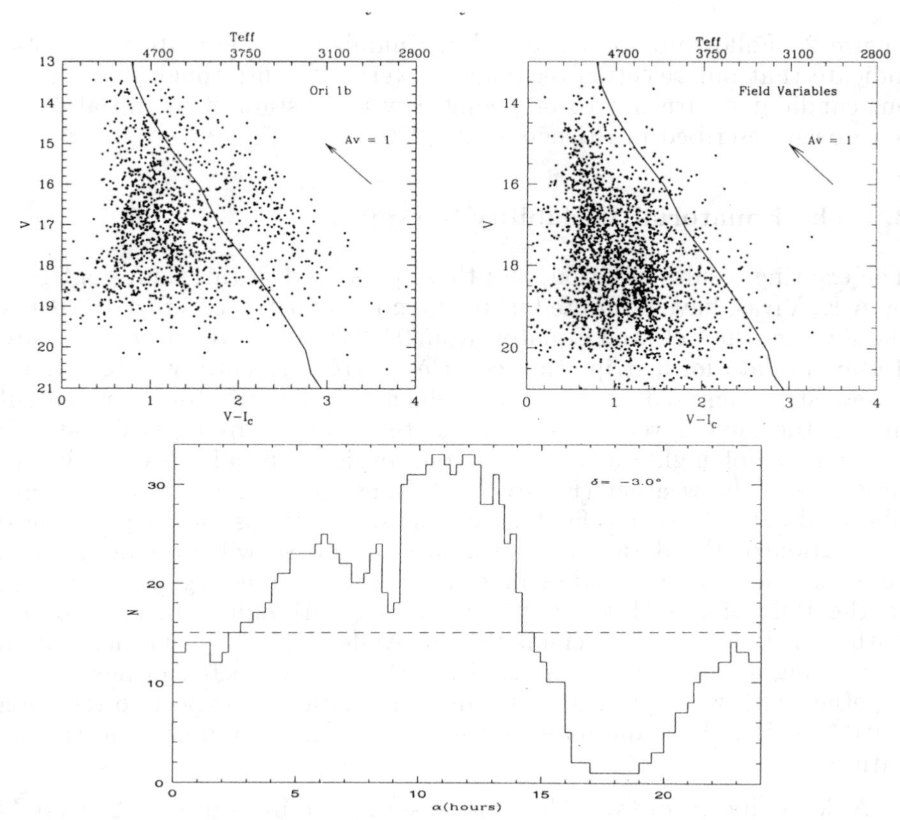

Figure 2. The top two panels show the V vs. $V - I_C$ diagram for all known variables in Ori 1b (left), and all variables in a control field offset from the Orion region (right). The solid line is the ZAMS. Note that the PMS variables in Ori 1b are clearly separated from the background, but they do not show up in the control field. The bottom panel shows the coverage in R.A. of the variability survey which is centered at $\delta = -3°$

1997) and the surroundings of the star σ Ori (Walter et al. 1998). A few earlier large scale studies (Wiramihardja et al. 1991; Alcalá et al. 1996) did not find the more dispersed stars that were slightly older and lower mass. Only by identifying and studying somewhat older ($\sim$ 10 Myr), low-mass stars in widely spread stellar populations can we address fundamental questions like the duration of the planet formation phase in dusty disks around young stars, and the lifetimes of molecular clouds.

We have developed tools for identifying variable stars using differential photometry. With a χ^2 test and assuming a Gaussian distribution for the errors, we consider as variable only those objects for which the probability that the observed distribution is a result of the random errors is $< 0.01\%$. The value of variability for picking out young stars is shown in

Figure 2. Follow-up spectra of a large number of candidate young stars indicate that our selection technique is very efficient: roughly 50–60% of our candidates turn out to be young, low mass stars. Our initial results have been described in Briceño et al. (2001).

3. The Equatorial Variability Survey

The exciting results in Orion, and those presented in this same conference by A.K. Vivas, have prompted us to extend our spatial coverage to span all the sky near the Celestial Equator, from 0^h–24^h. We have now accumulated 4 years of data for a strip centered at $\delta = -1°$. Presently we are scanning a new strip centered at $\delta = -3°$, which we plan to observe repeatedly during the next 2 years. The actual temporal sampling is dictated by the number of nights available for our project, usually 8–10 nights per month, and the weather (Figure 2c). Using the YALO telescope we have obtained data of selected fields along our survey strips for the photometric calibration of the data. The variability database will provide a unique research tool for our studies of nearby star forming regions, RR Lyrae in the Halo of the Milky Way, and quasar variability. In collaboration with E.A. Ponsot at Universidad de los Andes, Mérida, Venezuela, we are developing a variability database from the data collected in our surveys, together with web-based tools to mine the data. We expect to release an initial version of the database to the astronomical community in the near future.

Acknowledgments. This work is supported by grants S1–2001001144 from FONACYT, Venezuela, AST–9987367 from the National Science Foundation, and NASA Origins NGC5–10545.

References

Alcalá J.M. et al. 1996, A&AS, 199, 7

Akerlof, C. et al. 2000, AJ, 119, 1901

Baltay et al. 2002, PASP, 114, 780

Briceño, C. et al. 2001, Science, 291, 93

Carpenter, J.M., Hillenbrand, L.A. & Strutskie, M.F. 2001, AJ, 121, 3160

Brown, A.G.A., de Geus, E.J. & de Zeeuw, P.T. 1994, A&A, 289, 101

Herbst, W. 1986, PASP, 98, 1088

Hillenbrand, L. 1997, AJ, 113, 1733

Walter, F.M., Wolk, S.J. & Sherry, W., 1998, in Cool Stars 10, eds. R.A. Donahue and J.A. Bookbinder, ASP Conf. Ser., 154, 1793

Wiramihardja, S., Kogure, T., Yoshida, S., Ogura, K. & Nakano, M. 1993, PASJ, 45, 643

International Symposium on Astrophysics Research and on the Dialogue between Science and Religion, Vatican Observatory, 2003
C. D. Impey & C. E. Petry, editors

The Two Aspects of God

Francesca D'Antona
INAF, Osservatorio Astronomico di Roma, I–00040 Monteporzio, Italy

Abstract. I summarize how the conclusion of some philosophical approaches, that God has two different aspects (the first one necessary, the other one linked to contingency) has an interesting point of contact with the trinitary Christian faith, in which the Holy Spirit is the interacting Person. Taking this point of view, the inter-religious dialogue can be founded on the link through the "wind blowing where it pleases."

1. Introduction

I would like to explore in this article my views about the connection between religion and science, based on my personal beliefs and stimulated by the recommended readings for the Symposium.

One important thing we should remember as scientists is that common people look at us, and expect to find answers from people who know so much (or say so). Common people do not expect, or know, that we are all in the same boat with respect to their most difficult questions. There are three possible attitudes when we pose the problem of the connection between religion and science:

1. We may be interested in the intellectual understanding of whether a creator entity exists.

2. We may explore whether there is consistency between religious faith or faiths and the world as it is described by science.

3. We may be interested in trying to (intellectually) realize whether the possible entity God has a role in our life.

As we cannot tell people—based on science—whether God exists or not, the main question on which scientists can give an interesting word is about the consistency (or inconsistency) between science and religion. But the third question concerns what matters, and what is really interesting to people. In fact, a creator God who does not interact with his creatures may be of intellectual satisfaction, but he cannot change, or direct, or give hope to their lives.

In addition, it is not even interesting for scientists, as the premise of any good science is that we must look for contingent explanations to the nature of behavior, pushing non-natural explanations as far as possible, so that they do not interfere with our methodologies. It is obvious that understanding the motions of planets and stars and galaxies as result of the action of Newton's law, or as a consequence of general relativity, is intellectually far more satisfying than invoking the influence of angels to achieve the same result.

As science must necessarily limit its field to the contingent world, it seems that the ensemble of interactions between science and religion should be void. For that we should be glad as we can avoid the embarrassing situation of either having scientists who ground their atheism on "scientific" foundations, or of having God invoked as the cause of facts which, subsequently, are placidly explained by new developments of science. As George Coyne says, "Religious scientists should never be tempted to see the hand of God at the root of their scientific accomplishments." Nevertheless, when science touches the biggest of today's problems, its intersection with religion becomes a frontier, and it is very difficult to deal with the frontier problems without the risk of confusion. Take, for example, the difference between the "vacuum" full of physical laws, and thus not "empty," and the nothing from which (*ex nihilo*) God creates the World in the Judeo–Christian tradition (*creatio non est mutatio*). This frontier region is the main topic of this debate, as it has always been. We should try not to be confused by today's spectacular scientific achievements; the same feeling was held at the time of the Copernican revolution, at that of Darwin's *Origin of Species*, at the time of the discovery of the *Island Universes*, and so on.

2. The Dilemma of a Necessary God Creating a Contingent World

From my personal point of view, however, a point of interest is whether we can have any insight on the nature of God without mixing science and faith, by a more direct approach.

Creation *ex nihilo* presumes that a necessary God chooses to put into being a contingent universe. The Christian approach is that God creates the world by an act of "free will," but what "to choose" means here, or "why" a necessary God should create a world I am not able to discuss. Let me follow here schematically the argument by Keith Ward, as summarized by Paul Davies in the book *The Mind of God*. Either the acts of God are truly necessary—and so they could not have been different, there is no choice—or they are *free*, and thus arbitrary—there is nothing to tell how they should be. In other words, if God is the creator of a contingent world, he must be contingent; if God is necessary, everything which comes from him must be caused necessarily.

Is there a way out of this problem? We should think about God as both out and into the time: one side *necessary*, and the other one *contingent*. He exists necessarily, but he lets himself be modified by what he creates, as there is freedom in the creation. This is what Davies calls the "bipolar" God.

In this context, let me open a parenthesis and make a comment about Coyne's statement that, when He put the universe into being, "not even God would know the outcome with certainty." If this is taken to mean that not even God could know that the universe would have produced the human life after 14 Gyr, I would prefer to put this suggestion in a way less problematic to the Christian belief, for which Christ is the head of all creation:

> He is the image of the unseen God, the first-born of all creation.... He exists before all things and in him all things are hold together. (Col. 1,15;17; see also the first chapter of John's gospel)

To me, the *necessary* God indeed takes into his fingers every virtual photon, and knows exactly which side the dice will show when cast. The other pole of God interacts with the creation, or "works with the Universe." If indeed we could say that God did not know for sure that the outcome of Creation would be man, there would be an inconsistency between science and Christian religion—see the second point above.

3. The Bipolar God and the Holy Spirit

A "bipolar" God is then required to solve the paradox of necessity and contingency. But the part of God which interacts with the world has a name in the Judeo-Christian tradition: it has the characteristics which belong to the Holy Spirit, third person of the Trinity, the Spirit of God. Starting with the Genesis, which begins with the divine wind (the Jewish *ruah*, breath, or blow) sweeping on the waters, the action of the Spirit is present through the whole Bible, and it culminates with the first phases of the life of the Church. Any time God "talks" to his people, it is always through the Spirit. Prophets are plain men, apart from when they prophesy, under the influence of the Spirit. The same Jesus presents himself by reading "The Spirit of the Lord is on me" from the scroll of Isaiah in the synagogue, and ending with "This text is being fulfilled today." The Christian Church begins with the Pentecost and the action of the Spirit. I am always surprised when I think that the apostoles had seen Jesus resurrected, had talked with him, and what did they do? They went on with their fishermen work—hey had not understood anything at all, until the Pentecost.

4. The Spirit is the Link among Different Religions

Through the action of the Spirit we can reach a knowledge of God much better than through erudite studies of our scriptures (Christian or not), as it is well known to anybody who learns praying in the light of the Spirit.

The Spirit is the bridge between us and the mystery. The whole mystic tradition—both of the Oriental church through the so called "prayer of Jesus" or "prayer of the heart" of the Russian Pilgrim, and the Occidental tradition which culminates in the (incredibly modern) middle-age treatise *Cloud of Unknowing*—teaches how to empty the mind from everything (not only passions and sins, but even holy thoughts) in order to enter the mystery of God. Similar kind of prayers are also at the heart of many other religions; think, for example, about the "Shema Israel" of the Jewish tradition.

We are aware that "the wind blows where it pleases," (John, 3, 8) so any truth which is present in any religion should possibly be recognized as a revelation coming from the Spirit and establish links between the religious people of any doctrine.

5. Conclusions

In any discussion about these problems, one should get rid of pride and presumption, and let himself be guided by the Spirit. And, maybe our skills as scientists can be used to try to understand better how this interacting part of God interacts, and has interacted, with men and women throughout human history.

Acknowledgments. I wish to thank the organizers who gave me the opportunity to take part in this lively meeting, and George Coyne for his *lungimirante* activity in the project of the Vatican Schools and of the meetings dedicated to the Dialogue between Science and Religion. These pages would not have been written if Don Carmine Petrilli had not helped me, and many others, in the rewarding attempt to listen to the Spirit of God through prayer.

International Symposium on Astrophysics Research and on the Dialogue between Science and Religion, Vatican Observatory, 2003
C. D. Impey & C. E. Petry, editors

The Sun—Our Closest Star through the Eyes of *SOHO*

D. Dobrzycka

Harvard-Smithsonian Center for Astrophysics, 60 Garden Street, Cambridge MA 02138, USA

Abstract. The *Solar and Heliospheric Observatory* (*SOHO*) has been providing us with an unprecedented view of the Sun since 1996. The spacecraft carries twelve scientific instruments designed to study the internal structure of the Sun, its extensive outer atmosphere, and the origin of the solar wind. We present examples of *SOHO* multiwavelength observations of the solar corona and discuss some of the significant achievements related to our studies of heating and acceleration of solar wind, coronal mass ejections and solar jets.

1. Introduction

The *Solar and Heliospheric Observatory* (*SOHO*) is an advanced space telescope providing us with unprecedented view of the Sun in various ranges of the electromagnetic spectrum. It was launched in late 1995 and is operated from a permanent vantage point 1.5 million km sunward of the Earth, in a halo orbit around the L1 Lagrangian point. Such location of the spacecraft allows 24 hours per day continuous observations of the Sun. *SOHO* is designed to study the internal structure of the Sun, its extensive outer atmosphere and the origin of the solar wind. The spacecraft carries twelve scientific instruments, including three devoted to helioseismology, five to coronal, and four to heliospheric observations. Detailed information regarding mission and scientific instruments can be found at http://sohowww.nascom.nasa.gov. Here we present an example of *SOHO* multiwavelength observations of the solar corona that have revolutionized our understanding of the Sun and of the interactions between the Sun and the Earth's environment.

2. Ultraviolet Spectroscopy of the Solar Corona

The corona is the outermost layer of the solar atmosphere, right above the disk of the Sun. It eventually becomes the solar wind that expands into interplanetary and interstellar space.

Figure 1. Composite image from the *SOHO* instruments showing the UV light emitted by the solar corona at solar minimum. The solar disk image is from the EIT instrument and the extended solar corona (outside the black ring about 0.5 $R_\odot$ from the disk) is from the UVCS instrument.

Figure 1 shows the solar corona observed by *SOHO* in the ultraviolet part of the spectrum at the minimum of solar activity in 1996. The disk image was obtained by the *Extreme Ultraviolet Imaging Telescope* (EIT) in narrow spectral band centered at Fe XII (195 Å), while the extended corona is from the Ultraviolet Coronagraph Spectrometer (UVCS) and represents the total intensity emitted in the O VI (1032 Å) line. The dark regions above the poles and the narrow structure across the disk are the coronal holes, which connect with primarily open magnetic field lines. They have been identified as sources of the high speed solar wind that can reach velocities up to $\sim$ 750 km s^{-1}. The brighter structures in the equatorial region are streamers that contain denser plasma trapped in mostly closed magnetic field lines.

One of the most surprising results from UVCS/*SOHO* was that the heavy ions in the solar wind move much faster than electrons and protons (Kohl et al. 1998, Cranmer et al. 1999, see Figure 2). Another surprise was the extremely broad profile of O VI ions measured high in the polar coronal holes (Kohl et al. 1997). The widths of the coronal emission lines tell us about the ion velocity distribution along the line of sight. These broad O VI lines yield velocities up to 500 km s^{-1}, which corresponds to kinetic temperatures of 200 mln K. This is with contrast to the solar disk observations where the respective O VI lines are much narrower, with widths of about 30 km s^{-1}. So far, the best theoretical explanation is that the heavy ions (including the O^{5+} particles) absorb energy from high

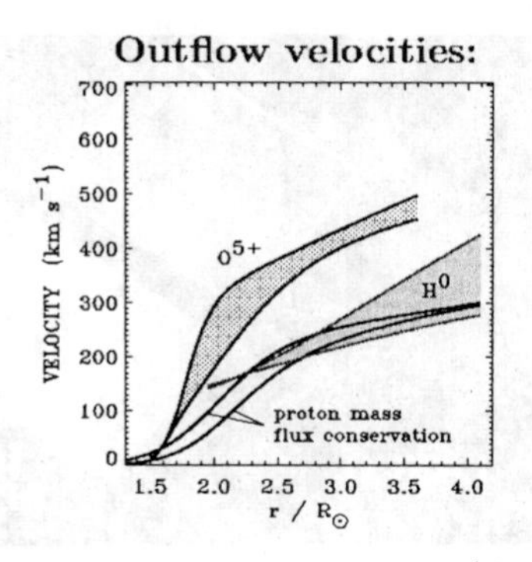

Figure 2. Empirically derived particle outflow speeds over the solar poles in late 1996 and early 1997. The shaded regions denote the range of hydrogen and ionized oxygen (O^{5+}) speeds able to reproduce observational data from UVCS. The lines denote the proton outflow speed derived from mass flux conservation: for a time-steady flow, the product of the density, speed, and flow-tube area should be constant, and density, polar flow-tube information, as well as *in situ* mass flux data, were used to define the constant. From Kohl et al. (1998).

frequency (short period) waves in the solar wind. These waves must have periods of about a millisecond for the particles to be able to sap their energy. In contrast, most of the observed waves in the solar wind have periods of more than a few seconds, up to minutes and hours. The UVCS observations thus suggest production of these shorter period waves in the solar corona (Kohl et al. 1998; Cranmer et al. 1999).

The *SOHO* coronal instruments provide unique observations of the dynamic solar phenomena, such as coronal mass ejections (CMEs), jets, flares, etc. Figure 3 shows a typical example of a CME as observed by the Large Angle and Spectrometric Coronagraph (LASCO) in white-light and UVCS in the UV part of the spectrum (Ciaravella et al. 2003). CMEs are ejections of mass and magnetic field from the lower corona into the higher solar atmosphere and interplanetary space. They involve large scale changes to the coronal structure and reconfiguration of the coronal magnetic field. Some of the CMEs hit the Earth's magnetosphere producing magnetic storms. Thanks to the *SOHO* coronal instruments we can follow CME's from their formation on the disk into further expansion in the higher corona. Imaging instruments tell us about morphology and velocity projected on the plane of the sky, while spectroscopy provides information about proton, electron, and ion temperatures, bulk 3D flows, elemental abundances, the ionization state of the plasma, and its helicity (the handedness and geometry of the twisted magnetic field).

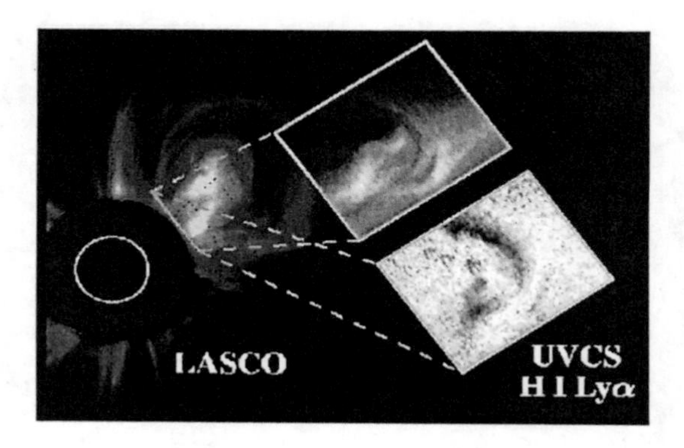

Figure 3. The coronal mass ejection observed by *SOHO* on 12 February 2000. The white-light image on the left was taken with the LASCO coronagraph. The position of the UVCS entrance slit at 2.3 $R_\odot$ is marked and an example of the UVCS results is shown on the right: time-lapse images of the intensity (top) and Doppler shift velocity (bottom) of the H I Lyα spectral line, as it evolved over a period of about 10 hours. From Ciaravella et al. (2003).

EUV jets ($2^\circ - 4^\circ$ wide) propagating high into corona were discovered by *SOHO* within the first months of the mission (Gurman et al. 1996). They appeared to originate near flaring UV bright points in prominent at solar minimum polar coronal holes. Analysis of the white-light and spectroscopic observations allowed to characterize jet dynamics and plasma conditions (e.g. Dobrzycka et al. 2000, 2002).

Acknowledgments. *SOHO* is a project of international cooperation between ESA and NASA. This work is supported by the National Aeronautics and Space Administration under grant NAG5–11420 to the Smithsonian Astrophysical Observatory, by Agenzia Spaziale Italiana, and by the ESA PRODEX program (Swiss contribution).

References

Ciaravella, A. et al. 2003, ApJ, to be submitted

Cranmer, S.R. et al. 1999, ApJ, 511, 481

Dobrzycka, D., Raymond, J.C. & Cranmer, S.R. 2000, ApJ, 538, 922

Dobrzycka, D. et al. 2002, ApJ, 565, 621

Gurman, J.B. et al. 1996, EOS, 77, 559

Kohl, J.L. et al. 1997, Sol. Phys., 175, 613

Kohl, J.L. et al. 1998, ApJ, 501, L127

International Symposium on Astrophysics Research and on the Dialogue between Science and Religion, Vatican Observatory, 2003
C. D. Impey & C. E. Petry, editors

How We Explain Things: The Status of the Laws of Nature

Witold Maciejewski

Osservatorio Astrof. di Arcetri, L.go Fermi 5, 50125 Firenze, Italy, and Obserwatorium Astronomiczne Uniw. Jagiellońskiego, Poland

Abstract. Laws of Nature play crucial explanatory role in contemporary science, although the concept of such laws comes from the epoch which commonly accepted a divine lawmaker. Using the example of contemporary physics, I examine what we now understand under the notion of the Law of Nature, and whether there is an alternative to the view that Laws of Nature are imposed externally on the world. I show that these laws can originate from within the world, but then they are merely a reflection of Order in Nature. This conclusion strikingly resembles the traditional philosophical view of Nature. Thus Order in Nature cannot be explained through laws originating from within the world.

1. Introduction

The main cognitive role of science is to explain things: why something is the way it is, how did it happen, where did it come from. In short, to explain something means to give its cause or reason. Science has attempted to answer these questions since its earliest days, when it still was one with philosophy. Various distinct types of explanation contribute to giving the reason to an event or a thing as completely as possible. For example: I hang my coat and it drops on the floor soon after that. Why did that happen? One thing is that there is gravity or that air closes above it, another thing is that the hanger was weak and it broke. These two reasons contribute to the explanation.

Aristotle sorted various causes into four groups, giving us the most complete theory of explanation in antiquity. He singled out four causes:

1. Material cause: the material of which the thing is made (like stone for a sculpture), but also its essential attributes (like the property of fire to rise up).

2. Formal cause: the shape of the thing (like the image of human body for a sculpture) but also its description (like notes for the music).

3. Efficient cause: maker or builder of the thing, something that creates it or makes it happen (like a sculptor for a monument or blowing wind for a falling tree).

4. Final cause: the purpose or function of the thing (like sitting for the chair, or house for the act of building).

This may look to us like an eclectic mix of things that we believe now to be causes of things or events, and of those that we don't. In particular, contemporary science rejects any teleology, and therefore the final cause does not apply when we explain nature, except for the directed actions of conscious beings. Moreover, contemporary science requires that we explain the world's phenomena within the world, thus all three other causes cannot include factors that transcend the world. Aside from these remarks, the listing of causes provided by Aristotle has proved complete enough, so that later theories of explanation limited themselves to stressing one or another cause without adding any new ones. Therefore in the next section, I relate contemporary scientific explanations to the Aristotelean causes, and try to locate the Laws of Nature in this context. In §3, I show that the concept of the Laws of Nature proposed by Newton contradicts the requirements of contemporary science listed above. In an attempt to find out where Laws of Nature come from, I examine closely in §4 how they function in the contemporary physics. Then I propose my interpretation in §5.

2. Scientific Explanation in Aristotelean Terms

It may be instructive to examine whether contemporary scientific explanation can be classified in terms of the Aristotelean causes. This explanation has usually two distinct aspects. For example the following sentence:

> *"There is volcanic activity on Jupiter's moon Io, because Io is located very close to Jupiter, and tidal forces generate friction, which heats up the interior of Io, and causes this activity."*

can be broken into two parts. First, there are the physical laws and their derivatives that shape the system (tidal forces of gravity, generation and transport of heat, etc.). Second, these laws would not have a chance to shape the system if Io were not so close to Jupiter. This second aspect of the explanation above belongs to the broad class of initial conditions. In this limited explanation, it is an accidental fact that Io moves close to Jupiter, as it is an accidental fact that a strong wind came and blew down a tree used as an example in the introduction. Thus we see that the aspect of explanation that involves initial conditions plays the same role as the efficient cause in the Aristotelean system.

Now, what is the role of the first aspect of explanation singled out above, i.e., of the Laws of Nature? In the ancient and medieval views, certain behaviour under given circumstances was perceived as an intrinsic property of the thing. To use a contemporary example, circular motion in a magnetic field defines an electron to be an electron. In this view, instead of the Laws of Nature we have the essence of things, and explanation in these terms refers to the Aristotelean material cause. In the example from the introduction, there is no physical law that tells the fire to rise up, but this is merely the property of the fire.

Whenever Laws of Nature are introduced, they subordinate things to themselves. In this view, the things behave the way they behave not because they are what they are, but because they follow the laws. A distinction appears between the law and the thing that follows it. The law describes what happens to the thing: it says how an event will look, regardless of whether it actually happens. This has a direct analogy with the example of the notes given in the introduction: notes describe the music that may or may not be played. Thus the Laws of Nature correspond to the Aristotelean formal cause.

3. How Laws of Nature Explain Things

In the previous section, I showed that current scientific explanation involves the Aristotelean efficient cause (initial conditions) and the formal cause (Laws of Nature). This second one takes the place of the material cause (essence of things) used in pre-scientific explanations. To what extent can Laws of Nature explain things? Consider a chair that falls apart when someone heavy sits on it. A law can be formulated here, derivative of the binding forces versus gravity, which states that at a certain point the chair cannot stand the weight. The chair follows this law, and falls apart. This sounds like a convincing explanation, and we accept it. Now consider motion of an electron in electromagnetic field, with electric and magnetic vectors perpendicular to each other. The electron moves on a helix. Why? Because it is subject to the Lorentz force

$$\frac{d\hat{\mathrm{p}}}{dt} = e\hat{\mathrm{E}} + \frac{e}{c}\,\hat{\mathrm{v}} \times \hat{\mathrm{H}}, \tag{1}$$

and if we calculate motion under this force, we get helical motion. There is something incredible in this explanation. Does the electron understand the Lorentz force? How does it interpret its various terms? These are complicated vector expressions — we understand them only because we are equipped with the intellect.

To Newton however, there was nothing strange in this explanation. His view rested on a religious conviction that there is an Architect, who designed this world. Things move the way they do in order to fulfill the Architect's plan. Thus, at their origin, the Laws of Nature were perceived as coming from outside of the world, and given by some external agent.

We may reject this view nowadays, but we are still left with an unsolved issue why things follow the laws, which are mathematical in structure, and accessible only to the intellect. It turns out that no matter what our convictions are, if we want such Laws to be *imposed* on Nature, then we imply that *mind comes before matter*, i.e., that there is an Intellect, which imposed the laws on the world.

Is there a way to have Laws of Nature as an explanation that does not involve mind, and that comes from within the world? In the next section I show that a more formalized physical approach creates this possibility.

4. Physical Space and the Realm of Mathematics

The example of an electron in the field was given in the physical space, where the electron is just a particle, a dot, and it moves according to the Lorentz force. In this approach, the law, which manifests itself in the expression for the force, is indeed imposed on the world. We can discover this law, but we cannot proceed with asking where it comes from. This is how far we get if we want to confine our explanations to within the world.

Working in the physical space involves mathematical formulae to describe motions of the electron, but the electron itself remains a particle from the realm of physics. However, physics can be formalized to the extent that mathematics is not only its language, but also its body. In other words, one can ascribe to the electron a mathematical formula, so that not electron but rather this formula is subject to investigation. Such a formula is called the action function, and rules governing the electron's behaviour can be derived from it. The action function for an electron can be written as

$$S = \int_A^B (-m\,c\,ds \;-\; \frac{e}{c}\,A_i\,dx^i). \tag{2}$$

This mathematical object is subject to a simple law

$$\delta S = 0, \tag{3}$$

which is the principle of least action. It can be shown that in general, once we assign mathematical objects of type (2) to a given physical system, the physical laws governing this system can be obtained by applying the principle of least action (3) to these mathematical objects. Thus the principle of least action emerges as a *fundamental law* from which particular laws follow. This law is so self-explanatory that it is likely to be intrinsic to the world, i.e., it does not need an external agent that imposes it on the world. It is reasonable to accept that it is just the property of the world that whatever happens in it, requires least action. We expressed here the principle of least action in the language of mathematics to accommodate sophisticated formulae of type (2) ascribed to the reality. However, this principle itself is of very primitive nature. Thus we have reached the goal stated at the end of §3: we have the Law of Nature that comes from within the world, and that does not involve mind.

Note however, that in the realm of mathematics, the electron is no longer a simple, featureless particle. It is now a well defined object, whose properties are formally expressed by (2). One can observe an interesting asymmetry:

physical space:	featureless particle	follows	an elaborated law,
realm of mathematics:	complex object	follows	a simple principle.

Let us focus on the question how we moved from the physical space to the realm of mathematics. The main issue here is how did we get the action function (2), which represents an electron as we claim. It does not come from direct measurements or experiments, but rather reflects certain symmetries and regularities that we believe take place in the world in which the electron moves. In general, if some physical system is described by the action function

$$S = \int_A^B dL, \tag{4}$$

then dL expresses mathematically regularities of the world in which this system functions. What further laws are derived from the least action principle entirely depends on dL. The more insightful we are into particular regularities or symmetries taking place in the world of events that we want to explain, the more complete laws we will obtain by applying the principle of least action to the mathematical objects of type (4) possessing these regularities. It is amazing, how various symmetries assumed in dL result in the well known laws of physics. If time is uniform, we get energy conservation. If space is uniform, momentum conservation pops out. Isotropic space leads to the conservation of angular momentum.

5. Proposed Interpretation and Consequences

Working in the realm of mathematics allowed us to base our scientific explanation of the behaviour of a system on the insightful description of regularities in this system, and on the simple universal law that is the principle of least action. This explanation does not include any factors that transcend the world. However, there is a price to pay for it: a lot of information about this world is taken for granted prior to giving the explanation. Moreover, this information enters in a particular form of symmetries and regularities that we observe or expect in the system. Regularities and symmetries mean *order*. Thus, as input to our scientific explanation of the world we already assume order in this world. And this order, which is formalized into the action function (4), gives all the amazing laws of physics through the principle of least action. Order means information encoded in the world. By observing certain regularities, we decode part of this information, and express it in the Laws of Nature.

Thus our explanation, in which laws of physics are intrinsic to the world, must assume that there is order in the world, and that science cannot provide an explanation for this order. Consequently, *Laws of Nature cannot explain Order in Nature.* If we want to get one, we must assume the other. If Laws of Nature always existed, then there was always order in the world, or information encoded in it, even if the world looked chaotic at times. Laws cannot create the world from nothing, as some scientists would like it to happen, because these laws themselves are manifestation of something that already exists, and that is intrinsically ordered. If we reject this view, we are back to the concept that laws are imposed on the world by a transcendental architect.

This finding has interesting consequences for the so called "fine tuning" of the Universe. We heard enough about it during this meeting, and I will refer here only to the notion that our world would be deprived of much of its complexity if some physical constants or laws were changed. Seekers of fine tuning change the gravitational constant or the electron charge by a bit, calculate the evolution of the world under such modified laws, and conclude that it all would be a mess. Given what we have shown in this paper, this is not a surprise. We have the laws that we do because our Universe is highly ordered. These, and not other laws, are manifestations of this order. If we change things in the laws, the new laws created in this way will no longer reflect this order. They will be just arbitrary mathematical expressions. Advocates of multiple universes may say that these new laws may describe different realities. This is not true either, because laws reflect the reality, and do not construct it. We do not have an access to a reality different from our Universe, therefore we are not able to discover laws governing it, and laws of our Universe are most likely irrelevant to such a reality.

6. Conclusions

In this paper I tried to find out how far one can proceed with the scientific explanation that does not go beyond the physical world. I showed that explaining things by the Laws of Nature does not necessarily imply a transcendental lawmaker. Laws of Nature can be obtained within the world, but for that, Order in the world has to be assumed. Thus Laws of Nature are manifestation of Order in Nature. This looks strikingly similar to the traditional philosophical approach. If Laws of Nature always existed, then Order is intrinsic to our Universe, but science cannot provide explanation for this Order.

International Symposium on Astrophysics Research and on the Dialogue between Science and Religion, Vatican Observatory, 2003
C. D. Impey & C. E. Petry, editors

El Observatorio Astrofísico Nacional de Tonantzintla: 60 Years of Excellence in Research

R. Mújica, M. Chavez and O. López

INAOE, Luis Enrique Erro No.1, Tonantzintla, 72840 Puebla, México

Abstract. We present a short overview of the past, present, and future of the Observatorio Astrofísico Nacional de Tonantzintla (OANTON), predecessor of INAOE our home institution, on the occasion of the 60th anniversary of the foundation of the OANTON.

1. Introduction

The year is 1942. On a hill that exposes the pre-Hispanic history of Mexico, some buildings and dispersed houses just constructed are seen, their yellow color shining intensely under a winter sun. The arrival of the president, Don Manuel Ávila Camacho, and his retinue, is expected. Around the stairs of the main building facing the two volcanos, the leaders of the town of Tonantzintla, the state and the city of Puebla, a representative group of recognized intellectuals, professors, and Mexican scientists, Secretaries of State, deputies, senators, among others, and a number of North American astronomers are congregated. In this way, the inauguration of the Observatorio Astrofísico de Tonantzintla was celebrated on February the 17th 1942. This act would set the standard and would mark the beginning of an epoch which still continues. From "The Dawn of Astrophysics in Mexico" (Pismis 1986)

The Observatorio Astrofísico Nacional de Tonantzintla was the result of the effort of Luis Enrique Erro (1897-1955) who was a professional engineer, an amateur astronomer, a true polymath, a gifted educator, a successful politician, and a philanthropist. He played a key role in drafting the Mexican higher education system as founder of National Polytechnical School, which would later become the National Polytechnical Institute (IPN). As Senator in the Mexican Parliament, he introduced reforms to the third article of the Mexican Constitution. This article proclaims that universal education is guaranteed as a right and should be provided by the State. He was also diplomat and served as ambassador in several countries.

As an astronomer he was the founder and director of the OANTON, he was also a member of the board of editors of the Astrophysical Journal.

Variable stars were his passion, and he was an active and leading member of the American Association of Variable Stars Observers (AAVSO), which gained him a close and fruitful friendship with Leon Campbell, the founder and coordinator of the association. Campbell would put Erro in contact with Harlow Shapley, Bart Bok and other important astronomers at the Harvard Observatory. These astronomers encouraged and helped Erro with the design, construction, and installation of a Schmidt type telescope in Tonantzintla. This telescope would later become the cornerstone of modern astronomy in Mexico.

2. The Early Years

Erro assembled a distinguished group of scientists and technicians at OANTON. Carlos Graeff Fernández, Paris Pismis, and Luis Rivera Terrazas, among others. They all made great efforts to make the Schmidt Camera operational. In the 1940's this telescope was one of the largest telescopes of its kind in the world. Important discoveries soon followed as a result of the ingenuity of the Mexican astronomers with access to state-of-the-art technology.

The first years at the OANTON were hard. After the initial euphoria, the Mexican astronomers had to find their own way, since the results were not immediate. Erro was faced with the lack of human resources to make full use of the Schmidt Camera. He therefore sent some people from the OANTON staff for training at the most important American observatories. Guillermo Haro was sent to Harvard Observatory to make studies on instrumentation, and Terrazas was sent to Yerkes Observatory. In 1945, Haro and Terrazas returned to the OANTON. At the same time a new objective prism was acquired with which research could be diversified.

Despite the difficulties, an invaluable collection of plates was obtained. During the inspection of these plates a great number of astronomical objects were discovered and catalogued. Some of the most important discoveries with the Schmidt Camera were: a new class of galactic emission nebulae called Herbig-Haro (HH) objects, a new class of irregular variable stars called flare stars (Haro 1968), and a new class of galaxies called blue compact galaxies (Haro 1956), as well as many planetary nebulae and quasars. From a list of blue objects identified by Iriarte and Chavira (1957), 15 turned out to be quasars, these were denoted by the abbreviation TON, followed by its number in the catalogue. From another catalogue elaborated by Haro and Luyten (1962) with plates obtained at the Palomar Observatory, 80 quasars resulted and were denoted PHL (Palomar-Haro-Luyten) objects. The intensive work at OANTON led to the discovery of new comets, novae, and supernovae. It also brought diverse awards and international recognition.

3. From Erro to Haro

Erro, following his principles, left the direction of the observatory in the hands of the younger generation in 1952. Erro's passion for astronomy was recorded in the commemorative plate that he asked to be put on the base of the first solar telescope mounted in Tonantzintla. The inscription on the plate read "This instrument was acquired with most of the money that President Lázaro Cárdenas gave me to cure my deafness. Be left in México like one more of the memories of this great Mexican." In recognition to the astronomical work of Erro, the IAU named a crater on the dark-side of the Moon in his honor.

Guillermo Haro (1913-1988), who since 1948 was director of the Observatorio de Tacubaya, which was operated by the Universidad Nacional Autónoma de México (UNAM), was appointed new director of the OANTON. This event put Haro in the privileged position to unify the objectives of Mexican astronomy for almost 20 years. During this time, several important projects were started. We cite in particular the redesign and construction of a new mounting for the Schmidt Camera, and the construction of a 1–meter telescope in Tonantzintla. When Tonantzintla began to suffer the impact of light pollution due to the proximity of the city of Puebla, Haro was quick to start the search for a site for a new national observatory. San Pedro Martir in Baja California was found the best suited place. Later UNAM would open and operate the Observatorio Astronomico Nacional.

Haro stepped-down from the directorships of UNAM's Institute of Astronomy and focused on revitalizing the OANTON. Haro recognized the importance of optics and electronics for astronomy. Hence, he was impelled to start a transformation of the OANTON into a National Research Institute for Astrophysics, Optics and Electronics. The project was presented to the, then, Mexican President of the time Luis Echeverria. The project was welcomed and supported by Echeverria, as he was starting a strong campaign of modernization reforms. On November 12, 1971, Echeverria decreed the creation of the Instituto Nacional de Astrofísica, Optica y Electrónica, or INAOE.

Haro was appointed as the director of the new Institute. He immediately convoked D. Malacara and A. Cornejo to start the Optics division, and de la Rosa, Noble, and Chandrasekhar to start the Electronics division. The Astrophysics division was integrated by the members of the OANTON: Haro, Iriarte, Chavira, and Graciela González. Haro was also aware of the importance of graduate studies in the Mexican province. INAOE opened a graduate program in optics in 1972, being the first of its kind in Mexico.

In 1973, Haro recommended the design and manufacture of the primary mirror for a 2.12m telescope for a new observatory. This ambitious project was aimed to keep the level of Mexican astronomy competitive. This was the first large project undertaken by the Optics division of INAOE. During the same year the graduate studies program in Electronics and the construction of a microelectronics lab began.

4. Into the Future

Haro left the directorship of INAOE in 1984. Iriarte temporarily replaced him. Jorge Ojeda assumed the direction of the Institute from 1984 to 1992. During this period, the 2.12m telescope was finished and was dedicated on September 8, 1987, amidst one of Mexico's worst economic crises. The new observatory was named Observatorio Astrofísico Guillermo Haro (OAGH), located in Sierra "La Mariquita," north of the town of Cananea, Sonora. Astronomy at INAOE was also suffering a crisis. Haro, Iriarte, and González had retired from the staff and only a handful of new astronomers had been hired. Ojeda started a new program for human resources development to revitalize the astronomy division, and the authors of this contribution were part of it.

In 1992, Alfonso Serrano was appointed the fourth director of INAOE. Serrano, following the steps into excellence led by Erro and Haro, involved INAOE in one of the most important scientific/technological projects in Latin America: design, construction, and operation of the Large Millimeter Telescope (LMT) / el Gran Telescopio Milimétrico (GTM). LMT/GTM is a USA-Mexico bi-national collaboration. The University of Massachusetts is the principal participating institution for the US, while INAOE is the lead for Mexico. The LMT is a 50m diameter millimeter-wave telescope designed for principal operation at wavelengths between 1 mm and 4 mm. The telescope is being built atop Sierra Negra (Altitude 4600 m) in the state of Puebla, Mexico. The telescope construction expected to be completed in 2004. Its great collecting area, as well as the projected accuracy of its surface, will allow the astronomers to carry out new projects in astronomy from planetary studies to cosmology (Carrasco & Carramiñana 2002).

INAOE is now a large research center in pure and applied sciences with more than 100 researchers and 300 grad students. José Guichard was appointed INAOE's director in 2001. He is continuing the construction of LMT/GTM. If Erro were to visit INAOE nowadays, he would not believe how powerful his dream has been. Sixty years later astronomy in Mexico is as vibrant as ever and is pushing the frontier of knowledge.

Acknowledgments. The authors, who all became VOSS alumni before they began graduate studies, would like to express their deepest gratitude to the Vatican Observatory, in particular to Father George Coyne (San Jorge) for continuous support and motivation and for his extraordinary efforts to propel astronomy in developing countries. The Vatican Observatory is now part of the history that we have described in this contribution. We are supported by CONACyT grants J32098-E, J32178-E, and 28506-E.

References

Carrasco, L. & Carramiñana A. 2002, Astrophys. & Space Sci. Lib., Vol. 274 (Kluwer Academic Publishers: Dordrecht), p. 283

Haro, G. 1968, Stars & Stellar Systems Ser., Vol. 7 (University of Chicago Press: Chicago), p. 141

Haro, G. 1956, AJ, 55, 72

Haro, G. & Luyten, G. 1962, Bol. Obs. Ton. y Tacub., 3, 37

Iriarte, B. & Chavira, E. 1957, Bol. Obs. Ton. y Tacub., 2, 3

Pismis, P. 1986, in Historia de la Astronomía en México, ed. M.A. Moreno, (FCE: México), p. 217

International Symposium on Astrophysics Research and on the Dialogue between Science and Religion, Vatican Observatory, 2003
C. D. Impey & C. E. Petry, editors

Can Science and Mysticism be Reconciled?

John T. Stocke

Center for Astrophysics & Space Astronomy, Box 389, University of Colorado, Boulder CO 80309-0389, USA

Abstract.

In the history of science, ever since the famous trial of Galileo, it has repeatedly been claimed that scientific truth cannot be reconciled with religious interpretations of the world. Now, at the beginning of the 21th century some individuals (e.g., Fritjof Capra, Gary Zukav) have claimed that a direct connection exists between the world view of modern physics and that of mysticism. Has modern science somehow proven or validated the mystical religious experience? Or does a basic and abiding conflict remain between these diverse human endeavors? A very brief summary of conclusions on these questions which have arisen from a course taught by the author will be presented. Also presented for the reader's use is an experiential exercise used in the course.

1. Discussion

For the last several years I have been facilitating small class discussions on the topic of science and mysticism at both the University of Colorado and Naropa University here in Boulder, Colorado, USA. The former is a public, state university, the latter a private, Buddhist university, which requires that all students have a meditation practice of some form. We begin each class with 20 to 30 minutes of vipassana meditation and/or guided imagery. Discussions of the readings and further experiential work follow in each class period.

I have listed the readings for the course in the references to this article. After reading both a general introduction to modern physics and Capra's seminal book, *Tao of Physics*, for historical and general background, we spend quite a bit of time reading and discussing the perspectives of Ken Wilber. An excellent introduction to his perspective on the relationship between physics and mysticism is in the forward to *Quantum Questions*. Wilber edited this book of mystical writings by last century's great physicists.

Central to Wilber's themes is the so-called Great Chain of Being from *The Perennial Philosophy* of Aldous Huxley. In Ken's rendition these are the realms of matter, life, mind, soul and spirit, with each successive level including and radically transcending the prior levels (e.g., the consciousness of individual minds is made up of living matter but somehow transcends these levels of reality). The dual nature of spirit is also recognized explictly in the Great Chain as both transcending all levels of reality as well as infusing the entire chain in an immanent or interpenetrating fashion. While different religious traditions often use different words and have different gradations of levels, it was Huxley's premise that all religious traditions incorporate the Great Chain of Being into their theology in some way. Science, on the other hand, works to reduce the higher levels of mind and life to their more basic constituents of matter, thereby discounting the potential existence of still higher levels of reality, which are acknowledged by all major religious disciplines.

2. An Exercise in Guided Imagery

Most of us do not have normal consciousness contact with the transcendent soul and spirit realms; nor do most of my students. I have found that the following guided imagery has on occasion proved helpful to introduce these very abstract concepts in a more accessible form. I invite you, the reader, to try the exercise yourself by having someone else read the individual indications to you, allowing the times in parentheses to elapse for each experience:

1. The exercise begins. Close your eyes, come into concentration on your breathing (3 minutes).
2. Visualize your brain as a dark void; empty of all substance of any description (2 minutes).
3. Now, visualize your brain as matter: atoms, molecules, protons, neutrons, electrons. And visualize your thoughts, including this visualization, as tiny electrical currents moving along the neural circuitry in your brain (3 minutes).
4. Now visualize your brain as life: see your brain as complex, self-replicating organic molecules; cells, neurons; see cells replicating these chemicals, replicating these cells. Visualize your thoughts, including this visualization, as changes and exchanges in the vital energies between cells. (3 minutes).
5. Now visualize your brain as mind. That part of you that says: "I am." The conscious, volitional mind, who thinks, who contemplates its own existence; who laughs; who cries; who loves and hates; who remembers; who gets depressed, and who gets elated. See your thoughts, including the current visualization, as the exclusive product of your mind (3 minutes).

6. Now visualize your brain as soul: the "watcher," the one with whom the mind holds conversation, the one who notices when the "I" feels angry or sad, happy or loving, and the one who is still there when your mind is blank, devoid of thoughts, devoid of tiny electrical currents, with no exchanges between cells and no individual volition. (3 minutes).

7. Now visualize the brain as spirit. That within each of us that responds with longing to the word "Namaste," which means "I honor that place in you; when you are in that place in you, and I am in that place in me, there is only one of us." That which is within and yet beyond yourself, beyond personality, beyond gender, thought and behavior, beyond mind and soul—the infinite and eternal in the finite you. (3 minutes).

8. Now return to concentrating on your breathing. Notice any thoughts that arise without attachment by simply noticing what they are; allow them to come to some completion, filling the space they leave behind with concentration on your breathing. (3 minutes).

Typically, I introduce this exercise incrementally to the students; i.e., we do the first three levels first, then add the soul and spirit levels in later class sessions. After each exercise, we leave plenty of time to relate and discuss experiences that come up during the exercise. I invite you to do the same with a friend or colleague.

3. Reconciling Two World Views

While the classes I have worked with on this material have not agreed upon whether modern physical science discoveries support a mystical world view, several problems with making these connections have been noticed (several by Wilber himself; see his Introduction to *Quantum Questions*):

1. Scientific theories change and therefore do not supply any indication of eternal truth, as required by mystical religious belief systems.

2. The approach of science is reductionistic, not holistic as embodied in the Great Chain of Being (see Arthur Koestler's definition of "holon," an object which is more than the sum of its parts). I have heard Huston Smith say it best: "From the perspective of science, we are the more that comes from the less. Whereas from the religious perspective, we are the less that are derived from the more." This is the essence of the Great Chain of Being view of reality, and is opposite to the views held by modern science.

3. These endeavors to prove or support mystical beliefs with modern science or to look to modern science for proof of mystical or religious beliefs, are a new type of "scientism;" i.e., a suggestion that the scientific method is the only viable method for obtaining truth, requiring all mystical belief to have scientific support of one form or another.

4. Finally, to take the perspective that science supports or validates mystical religious belief also suggests a belief system by which we can all live better lives and fulfill our life's purpose simply by adopting a new cosmology. To the contrary, all religious traditions have developed their own forms or "technologies" to develop spiritual understandings, that do not conform to the strict methodologies of science, nor can they be judged by scientific criteria (e.g., meditation, prayer, fasting, devotional works, ritual, non-ordinary consciousness experiences).

While the first three objections above may be viewed as primarily academic argumentation, the final objection takes on real world consequences for all of us as it suggests, even requires, that actions beyond science and the processes of our thinking minds, are necessary to fully incorporate spirit (of whatever form) into our lives. For astronomers this does not mean that we cannot find spiritual fulfillment in our work, but suggests that if we do, it will not be because of the scientific results that we achieve (e.g., any new cosmology that we might help develop) but rather by the manner in which we conduct our lives as we do our work. Indeed, the very same could be said for any profession.

Acknowledgments. I thank all my University of Colorado and Naropa students, Perk Clark, Jean McCreedy, Stan Grof, Ken Wilber, and Jeremy Hayward for helpful and inspirational conversations.

References

Capra, F. 1976, The Tao of Physics (Shambhala: Boulder)

Huxley, A. 1944, The Perennial Philosophy (Harper & Row: New York)

Weber, R. 1986, The Search for Unity: Dialogues with Scientists and Sages (Routledge & Kegan Paul: New York)

Wilber, K. 1990, Eye To Eye, The Quest for the New Paradigm (Shambhala: Boston)

Wilber, K. 1984, Quantum Questions (Shambhala: Boston)

Wolf, F.A. 1989, Taking the Quantum Leap (Harper & Row: New York)

International Symposium on Astrophysics Research and on the Dialogue between Science and Religion, Vatican Observatory, 2003
C. D. Impey & C. E. Petry, editors

The Interaction between Science and Religion in Post-Socialist Slovakia: A Personal Reflection

Juraj Tóth

Železničná 5, 900 31 Stupava, Slovak Republic

Abstract. The change of the state's materialist ideology in post-socialism countries brings a new dialogue between science and religion. What does this dialogue look like? Can it bring something helpful into the lives of people in such countries? The personal experiences of a young astronomer and believer, who lives in Slovakia, are presented.

1. Introduction

I would like to mention a few things about the dialogue between science and religion in my home country, Slovakia, one of two former countries of Czechoslovakia in middle Europe. I present my own view as a person who grew up under the doctrine of state atheism and later became a Christian by reading the Gospel. I hope that my experiences will be helpful in understanding the thoughts, life-styles and decisions of people who live in my country, where official atheism has been present for most of the last 50 years and Christianity has been practiced for more than 1000 years.

2. Brief History and Current Situation

What are the special characteristics of this region in middle and eastern Europe after the Second World War? Most countries had a socialist or communist regime with the state's doctrine based on the materialistic view of the world. The usual phrase from that time was: only materialism is the scientifically proven and, hence right, opinion. Everybody was required to embrace this opinion, otherwise he or she was persecuted. For example, religious persons and scholars were persecuted, as were children and young people who attended church. Furthermore, they had problems when they sought to continue their education.

What is the traditional religion in Slovakia? It is Christianity. Let me give a brief history, which is necessary to understand this background. The old Slavs came to this region in the 5th and 6th centuries. They accepted Christianity in about the 9th century after the mission of Saints Cyril and

Methodius, Greek brothers, who were asked to come by Moravian King Rastislav. For the next 1000 years, Slovakia belonged to the Hungarian kingdom where Christianity was also accepted. After World War I, the Slovaks joined with the Czechs to form a new state, Czechoslovakia, which endured until after World War II, but with a communist regime from 1948 until 1989. In 1993, Czechoslovakia split apart in a peaceful way into two republics: the Czech Republic and the Slovak Republic.

3. Dialogue

Let us take a closer look at the society in Slovakia after 1989, when a new political atmosphere came about. The Church was one of the winners after 50 years of persecution. But people were afraid that new political or religious doctrines would prevent them from being free. There is also still a mixing of lies and myths about Christianity—holdovers from the past 50 years. One example is the Galileo case: many people think that he was burned at the stake. Probably this myth came about from the fact that Galileo case was always mentioned along with Giordano Bruno in every textbook of physics at the primary and high school levels. Why did students learn about Giordano Bruno in physics? Simply, because it was part of the State's propaganda.

Despite this inhospitable climate for religious freedom, more than 80% of the citizens claim to be Christians (July 2002 census), but only about 20% are practicing their faith. Generally we can distinguish four different variations on the relationship between science and religion:

1. Those who think that the natural sciences and religion are in opposition with or in contradiction to each other. This is the most widely-held belief.

2. The creationists who use the Holy Scripture as a textbook of the natural sciences.

3. The believers in the absolute power of science, who think that the only way to knowledge is through the natural sciences.

4. Those who reject both science and religion.

Recently, it has been Christians who have opened a public dialogue about the relationship between science and religion, e.g., in their radio or TV programs. An adequate dialogue between between science and religion requires an acceptance of the goodness both of science and faith.

Acknowledgments. The author is very grateful to the Vatican Observatory for its kind support which enabled his participation in the second SVOSS.

International Symposium on Astrophysics Research and on the Dialogue between Science and Religion, Vatican Observatory, 2003
C. D. Impey & C. E. Petry, editors

The Leonid Meteor Shower: A New View?

Juraj Tóth

Astronomical Institute, Comenius University, Mlynska dolina, 842 48 Bratislava, Slovak Republic

Abstract. Observations and scientific progress from the monitoring of Leonid meteor shower activity from 1998 until 2002 are discussed. Also presented are the unique photographic observations of the 1998 shower which were performed at the Modra Observatory (Slovakia), and high quality HD TV observations of the Leonid storm in 1999 within the frame of the Leonid MAC campaign.

1. Introduction

The Leonids are meteors, sometimes referred to as a shooting stars, which annually display in the middle of November. They appear to radiate out of the constellation Leo that is why they are called Leonids.

The Leonid meteor shower has been known for more than 1100 years and is well recorded in the chronicles of the Asian, European and Arabic cultures. The first written records of observations of stars which fell like rain are from the 10th century. The meteor rains or storms are typical for Leonids and were repeated roughly every 33 years. However, only the storms in 1799 and 1833 drew attention of modern natural sciences. Olmsted (1834) and Twining (1834) established the notion of a parallel stream of moving meteoroids intersecting the Earth. Newton (1864) showed that stream had a sidereal period, so that these streams were interplanetary in nature. In 1865, the parent comet 55P/Tempel–Tuttle was discovered, and the anticipated and observed storm of the Leonids in 1866 provided the direct link between the orbit of a comet and a meteoroid stream. The subsequent return of the parent comet in 1899 and again in 1933 did not provide Leonids storms due to the dynamics of the comet.

Fred Whipple suggested in 1950 that meteoroids or dust particles can be ejected from a cometary nucleus by the drag of water vapor (Whipple 1951), and a meteoroid stream can be created by different ejection velocities and radiation forces (Plavec 1955, Kresak 1976). Such dust trails were first observed in 1983 by the IRAS satellite in the orbits of several short-period comets.

The comet 55P/Tempel–Tuttle was observed again in 1965 and in 1966, which was the greatest meteor storm ever recorded with a maximum frequency of about 150,000 meteors per hour from the North America in the late night of November 17. At that time about 40 meteors per second were seen in the sky. A detailed study of the comet dynamics and historical Leonid accounts showed that dust concentrates mainly behind the comet and just outside of the comet orbit (Yeomans 1981). From this it could be inferred that the Earth would cross the dust cloud in the comet vicinity, which last returned in 1998.

The return near the end of second millennium was for many people probably the last chance to see a Leonid meteor storm. In 1991 the International Leonid Watch (ILW) campaign was established to monitor the activity of the Leonid meteor shower in its entirety. These are the main goals of the program: to study the space distribution of meteoroids in the stream, the mass distribution of the meteoroids, the spectra of the Leonids and their trails in the atmosphere, and the fragmentation of the meteoroids, as well as to study the atmosphere and the ionosphere during the activity of the shower, and to document the collection of micro-meteorites in the high atmosphere. Visual observations worldwide were coordinated by the International Meteor Organization (IMO). During the next years the activity and also the expectations increased.

The parent comet, 55P/Tempel–Tuttle, returned to perihelion on February 28 of 1998. Its visual magnitude reached +7.7. The diameter of the comet nucleus was 3.6 km (Hainaut el al. 1998) and the rotational period was 15.33 ± 0.02 hours (Jorda et al. 1998).

2. The 1998 Leonids

The peak of the shower in 1998 was anticipated around the time of the nodal passage of the comet on November 17 at 19:43 UT (Yeomans et al. 1996). Many other authors predicted the peak of the main activity at that time within 2 hours (Jeniskenns 1996, Toth 1999, etc.), which implied that the best location for the observation would be in the western Pacific or in eastern Asia. Only Kresak (1993) predicted the maximum earlier on November 17 at 8:30 UT. These modeled predictions, based on historical and recent observation of the Leonid meteor display, were supported by the fact that the six largest storms all peaked 0.5 to 2 hours after the nodal longitude of the parent comet (Brown 1999). Predictions of shower frequencies were uncertain. The Zenithal Hourly Rate (ZHR) at maximum was predicted to be in a range from about 200 to 10,000.

That is why many scientific expeditions moved to Mongolia, China, Japan and to other eastern Asia countries to see the Leonids storm in 1998. Also, the first international Leonid Multi-Instrument Aircraft Campaign (MAC) coordinated by Dr. Peter Jenniskens and sponsored by NASA as the Astrobiology mission was executed over Okinawa, Japan. Their objectives were to learn how extraterrestrial materials may have been brought to the Earth at the time of the origin of life. They used new techniques of

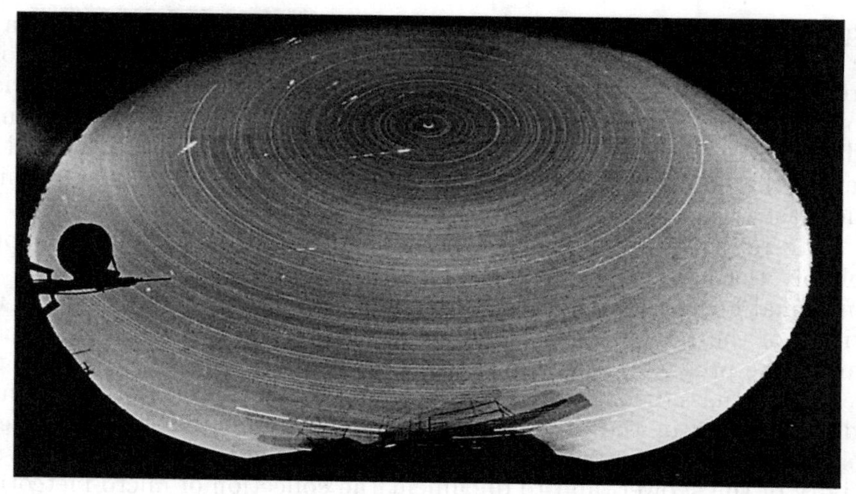

Figure 1. The fixed mode photograph of Leonids 1998 November 16/17 from Modra Observatory, Slovakia. The exposure last more than 11 hours. About 80 Leonids were recorded.

image intensified video and CCD imaging for meteor observations in new spectral windows (infrared, ultraviolet). Koschny and Zender (2000) have shown that observations from airplanes at high altitudes provide about a five times larger number of observed meteors compared to ground observation, due to lower atmospheric extinction near the horizon at high altitudes around 11 km. This is one of the main advantages of such airborne observation missions.

Arlt and Brown (1999) analyzed visual observations around the world in the frame of ILW. They obtained the time of maximum over eastern Asia which occured close to the predictions at 20:33 UT November 17 with the ZHR $= 136 \pm 5$. Many observers were disappointed because they had expected a higher frequency. But about 19 hours earlier the Leonids displayed an unexpected maximum with extraordinary large number of very bright fireballs over the western Asia and Europe at 1:55 UT with a maximum of ZHR $= 357 \pm 11$. The author of this paper was lucky to see these fireballs from the Modra Observatory, Slovakia. His estimation of the ZHR was 400.

We obtained only limited information from our single station photographic observation. We were able to identify 60 Leonids on both fixed and guided plates and to calculate their times of appearance from transformation of spherical coordinates, using the method for all-sky photographs developed by Borovicka et al.(1995). After that we studied the time distribution from which the maximum of the activity for bright Leonids at 1:40 UT November 17 (Toth et al. 2000) could be estimated. The relatively precise astrometry measurement (RMS = 1 arcmin) of the begin-

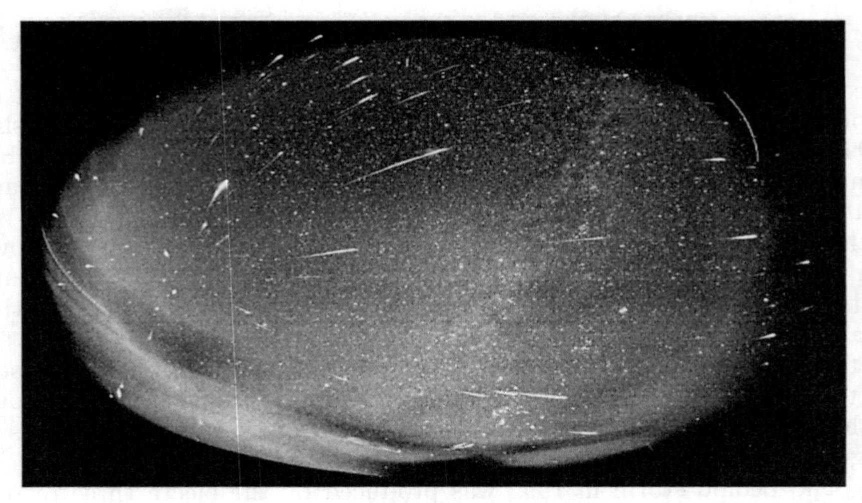

Figure 2. The guided mode all-sky photograph of Leonids 1998 November 17 from Modra Observatory, Slovakia. The exposure last for 4 hours. 156 Leonids were recorded.

nings and ends of the recorded meteor trails allowed the calculation of an average radiant position for the time of maximum: $\alpha = 153.29° \pm 0.05°$ and $\delta = 22.07° \pm 0.05°$ (solar longitude $234.52°$, 2000.0). In the next step we measured the brightness of meteors according to photometric procedure used by Ceplecha (1987) and obtained the absolute magnitude distribution. We found the minimum in the magnitude distribution one hour before maximum at 0:30 UT November 17, which means that the Earth encountered the largest meteoroids one hour before the maximum activity and that particles within that stream were not distributed randomly.

As we now know, meteor storms are produced by highly concentrated young filaments from each comet perihelion passage. The comet's orbit is still perturbed by gravitational and non-gravitational effects; each perihelion passage is slightly different. This difference is responsible for such distinct filaments which were studied for the first time by Kondrat'eva and Reznikov (1985). After the unexpected bright Leonid outburst in 1998 Asher studied filaments structure in details. He showed (Asher et al. 1999) that the outburst of bright Leonids in 1998 were produced by the filament ejected from the parent comet in 1333, which was in an orbital resonance 5/14 with Jupiter.

Spurny et al. (2000) reported the observation of extreme beginning heights of bright Leonids up to 200 km altitude and diffuse structures of meteors (comet-like) above 130 km. This was proven also for the Perseids (Koten et al. 2001). The physical mechanism responsible for jet-like ablation features and high altitude beginnings of meteors is not yet understood.

3. The 1999 Leonids

Predictions for the 1999 Leonids possible storm activity were based on papers by Kondrat'eva and Reznikov (1985) and McNaught and Asher (1999). The time of the peak activity was predicted at 2:08 UT ± 10 min. November 18. The best observation occured from western Asia and Europe. The observed maximum occurred at 2:02 UT $\pm$2min. Nov. 18 with ZHR $= 3700 \pm 100$. A second Leonid MAC Campaign was held over the Mediterranean Sea on the night of November 17/18. Dr. H. Yano and Dr. S. Abe from ISAS Japan carried out intensified video observation in a cooperation with NHK Japanese broadcasting company—one of the best video records of a meteor shower ever done. The result from just a part of the whole record showed a non-random distribution of meteoroids within the stream. This non-random distribution is due to a progressive fragmentation of meteoroids in space (Toth et al. 2002).

The Leonid storm in 1999 was produced by the ejecta three revolutions old, which means 100 years from the ejection from the parent comet (McNaught and Asher 1999). We found that only a fraction of the whole population was in groups. This result was later supported by Watanabe et al. (2002) who observed several groups of Leonids in 2001. They suggested a thermal heating as a process which leads to secondary fragmentation of meteoroids in space after the ejection from the parent comet.

4. The 2000 Leonids

In 2000, no Leonid meteor storm was observed, as was also predicted. That is why no official aircraft campaign was organized. The time of the maximum from visual observations around the world was 7:12 UT November 18 with ZHR $= 480 \pm 20$ (Arlt and Gyssens, 2000). That year was a good break year between Leonid storms for making analyses of the previous data, allowing us to prepare more thoroughly for the years 2001 and 2002 when the next storms were anticipated.

5. The 2001 Leonids

The theory of meteor stream evolution has significantly advanced during the Leonid activity since 1998. A new prediction by Lyytinen et al. (2001) was the closest to the observed maximum of the Leonid storms in 2001. This model assumes that particles leaving the comet suffer from solar radiation pressure which increases the semi-major axes of orbits. On the other hand previous models of Kondrat'eva and Reznikov (1985) and McNaught and Asher (1999) assume some ejection velocity in the forward direction compared to the comet motion. The mean peaks at 10:39 UT ± 4 minutes (over America with the ZHR $= 1620 \pm 40$) and at 18:16 UT ± 4 minutes (over Asia with the ZHR $= 3430 \pm 90$) match the predicted times of the 7 revolution and 4 revolution old dust filaments (Arlt et al. 2001). While the first maximum occurred 11 minutes after the prediction of Lyytinen et al. (2001) and 44 minutes after the prediction of McNaught and Asher (2001), the second maximum occurred 3 minutes after the prediction of McNaught and Asher (2001) and 4 minutes before the prediction of Lyytinen et al. (2001), which is within the uncertainty interval.

Watanabe et al. (2002) observed in 2001 during second maximum of Leonid storm from Japan several short-duration outbursts in which more than 30 to 40 meteors appeared in a few seconds. Their observation of 4 revolution relatively young filament meteors is an example of the recent fragmentation in space.

6. The 2002 Leonids

Two very strong activity peaks of the Leonid storm were predicted for November 19, 2002. The first one was expected over Europe and North Africa between 3:48 to 4:04 UT. The second one was expected over North and Central America between 10:23 to 10:47 UT. The actual first maximum occurred according to the first analysis (Arlt et al., 2002) at 4:10 ± 1 minutes UT with ZHR $= 2510 \pm 60$ and the second maximum at 10:47 ± 1 minutes UT with ZHR $= 2940 \pm 210$. The prediction of Vaubaillon (2002) (4:04 UT and 10:47 UT), based on theoretical model of numerous dynamically diverse meteoroids, was the closest to the observations.

Also, the author of this paper observed the first maximum of the Leonid storm 2002 from Modra Observatory, Slovakia, together with colleagues. We used various observation techniques, e.g., visual, photographic, TV, telescopic and radar observations. The time of maximum from our visual observation of one minute counts occurred at 4:10 UT with ZHR about 3000 ± 400 according to preliminary analysis. Despite a strong Moon light and increasing cloud formation around the time of maximum, this last Leonid storm was a marvelous farewell from the Leonid meteor shower until the end of this century, when we expect storm level activity again in 2098.

7. Summary

This brief description of the Leonid meteor shower activity is not complete. I should also mention other scientific discoveries and progress in this field: observations of the Leonid meteor shower from space, air and ground at different electromagnetic wavelengths; the first observation of impacts of large Leonids on the dark side of the Moon by video techniques in 1999 (Bellot Rubio et al. 2000); the detection of the Moon's sodium atmosphere during the Leonids in 1998 (e.g., Smith et al. 1999); theoretical investigations of meteor vapor temperatures and emission spectra (Popovova et al.

2000) and their relation to the interplanetary dust (Rietmeier 2000); the development of systems for early warning to satellite operators to minimize risk of meteoroid's impacts; the physical structure, grain size and mode of ablation of meteoroids indicative of a dust ball model (Murray et al. 1999, 2000); collection of many precise photographic and video orbits of Leonids from different filaments (e.g., Betlem et al. 2000); the first observation of dust trail of Leonids debris in space from the ground (Nakamura et al. 2000); the detection of ELF/VLF of faint Leonids (Price and Blum 2000); studies of interaction of meteor debris (ions and neutrals) with the ionosphere; and, studies of atmospheric meteor trains as a chemi-luminescence of metals (Borovicka and Jenniskens 2000). In a few years we will be able to evaluate all the knowledge we obtained during the Leonid return of 1998 to 2002.

Acknowledgments. This work was supported by the Slovak Grant Agency VEGA (grant No.1/7157/20). The author is very grateful to the Vatican Observatory for its kind support which enabled his participation in the second SVOSS, and to the editors for their special patience.

References

Arlt, R. & Brown P. 1999, WGN, 27, 267

Arlt, R. & Gyssens, M. 2000, WGN, 28, 195

Arlt, R., Kac, J., Krumov, V., Buchmann, A. & Verbert J. 2001, WGN, 29, 187

Arlt, R., Krumov, V., Buchmann, A., Kac, J. & Verbert, J. 2002, WGN, 30, 205

Asher, D.J., Bailey, M.E. & Emel'yanenko, V.V. 1999, MNRAS, 304, L53

Bellot Rubio, L.R., Ortiz, J.L. & Sada, P. V. 2000, Earth, Moon and Planets, 82-83, 575

Betlem, H. et al. 2000, Earth, Moon and Planets, 82-83, 277

Borovicka, J. & Jenniskens, P. 2000, Earth, Moon and Planets, 82-83, 399

Borovicka, J., Spurny, P. & Keclikova, J. 1995, A&A, 112, 173

Brown, P. 1999, Icarus, 138, 287

Ceplecha, Z. 1987, Bull. Astron. Ins. Czech., 38, 222

Hainaut, O.R., Meech, K.J., Boehnhardt, H. & West, R.M. 1998, A&A, 333, 746

Jenniskens, P. 1996, Meteorit. Planet. Sci., 31, 177

Jorda, L., Lecacheux, J., Colas, F., Frappa, E. & Laques, P. 1998, in IAUC 6816, ed., D.W.E. Green (Cambridge: Massachusetts)

Kondrat'eva, E.D. & Reznikov, E.A. 1985, Sol. Syst. Res., 19, 96

Koschny, D. & Zender, J. 2000, Earth, Moon and Planets, 82-83, 209

Koten, P., Spurny, P., Borovicka, J. & Stork, R. 2001, in Meteoroids 2001, ed., B. Warmbein (ESA SP-495: Kiruna, Sweden) p. 119

Kresak, L. 1976, Bull. Astron. Ins. Czech., 27, 35

Kresak, L. 1993, in Meteoroids and their Parent Bodies, eds., J. Stohl & I.P. Williams, (Astron. Inst., Slovak Acad. Sci.: Bratislava) p. 147

Lyytinen, E., Nissinen, M. & van Flandern, T. 2001, WGN, 29, 110

McNaught, R.H. & Asher, D.J. 1999, WGN, 27, 85

McNaught, R.N. & Asher, D.J. 2001, WGN, 29, 156

Murray, I.S., Beech, M., Taylor, M.J., Jenniskens, P. & Hawkes, R.L. 2000, Earth, Moon and Planets, 82-83, 351

Murray, I.S., Hawkes, R.L. & Jenniskens, P. 1999, Meteorit. Planet. Sci., 34, 949

Nakamura, R., Fujii, Y., Ishiguro, M., Morishige, K., Yokogawa, S., Jenniskens, P. & Mukai, T. 2000, ApJ, 540, 1172

Newton, H.A. 1864, Am. J. Sci., 38, 53

Olmstead, D. 1834, Am. J. Sci. Arts, 25, 354

Plavec, M. 1955, in Meteors, Special Supplement 2, J. Atmos. Terr. Phys., p. 168

Popova, O.P., Sidneva, S.N., Shuvalov, V.V. & Strelkov, A.S. 2000, Earth, Moon and Planets, 82-83, 109

Price, C. & Blum, M. 2000, Earth, Moon and Planets, 82-83, 545

Rietmeijer, F.J.M. 2000, Meteorit. Planet. Sci., 35, 1025

Smith, S.M., Wilson, J.K., Baumgardner, J. & Mendillo, M. 1999, Geophys. Res. Lett., 26, 1649

Spurny, P., Betlem, H., Jobse, K., Koten, P. & van't Leven, J. 2000, Meteorit. Planet. Sci., 35, 1109

Toth, J. 1999, in Meteoroids 1998, ed., V. Porubcan (Astron. Inst., Slovak Acad. Sci.: Bratislava) p. 223

Toth, J., Kornos, L. & Porubcan, V. 2000, Earth, Moon and Planets, 82-83, 285

Toth, J., Yano, H. & Porubcan, V. 2002, ISAS Report SP No. 16, in press

Twining, A.C. 1834, Am. J. Sci., 26, 320

Watanabe, J., Sekiguchi, T., Shikura, M., Naito, S. & Abe, S. 2002, PASJ, 54, L23

Whipple, F.L. 1951, ApJ, 113, 464

Yeomans, D.K. 1981, Icarus, 47, 492

Yeomans, D.K., Yau, K.C. & Weismann, P.R. 1996, Icarus, 124, 407

Distribution by Continents and Nations

SUMMARY: 9 Schools, 225 Students, 52 Nations
43% F, 57% M
60% from Developing
40% from Industrialized

Continent	Nation	Total
AFRICA:	Nigeria 4, South Africa 6	10
NORTH AMERICA:	Canada 10, Cuba 1, Mexico 8, United States 22	41
SOUTH AMERICA:	Argentina 20, Bolivia 1, Brazil 10, Chile 4, Peru 1, Uruguay 1, Venezuela 9	46
CENTRAL AMERICA:	Honduras 1	1
ASIA:	China 6, India 10, Indonesia 1, Japan 1, Korea 3, Sri Lanka 4, Taiwan 2, Thailand 1, Vietnam 1	29
EAST EUROPE:	Armenia 2, Bulgaria 7, Croatia 6, Estonia 2, Hungary 1, Iran 1, Lithuania 2, Poland 5, Romania 3, Russia 5, Slovenia 1, Ukraine 6, Yugoslavia 2	43
WEST EUROPE:	Austria 2, Belgium 3, Denmark 3, Finland 4, France 1, Germany 4, Greece 7, Italy 8, Netherlands 2, Portugal 1, Spain 3, Sweden 2, Switzerland 2, Turkey 4, United Kingdom 2	48
OCEANA:	New Zealand 7	7

Distribution by Continents and Nations

SUMMARY: [illegible] Students, [illegible] Nations
[illegible] M
[illegible]% from Developing [illegible]
[illegible]% from Industrialized [illegible]

Continent	Nation	Total
AFRICA:	Nigeria 4, South Africa [illegible]	[illegible]
NORTH AMERICA:	Canada 10, [illegible] Mexico [illegible], United States 22	[illegible]
SOUTH AMERICA:	Argentina 20, Bolivia 1, Brazil 10, Chile [illegible], Peru [illegible], Uruguay 1, Venezuela [illegible]	[illegible]
CENTRAL AMERICA:	Honduras 1	1
ASIA:	China [illegible], India [illegible], Indonesia 1, Japan [illegible], Korea [illegible], Sri Lanka [illegible], Taiwan 2, Thailand [illegible], Vietnam [illegible]	[illegible]
EAST EUROPE:	Armenia 2, Bulgaria [illegible], Croatia [illegible], Estonia [illegible], Hungary 1, Iran 1, Lithuania 2, Poland [illegible], Romania 3, Russia [illegible], Slovenia 1, Ukraine [illegible], Yugoslavia [illegible]	[illegible]
WEST EUROPE:	Austria 2, Belgium [illegible], Denmark 3, Finland 4, France 1, Germany 4, Greece [illegible], Italy [illegible], Netherlands 2, Portugal 1, Spain 3, Sweden 2, Switzerland 2, Turkey 1, United Kingdom [illegible]	[illegible]
OCEANIA:	New Zealand [illegible]	[illegible]

VOSS Alumni, 1986–2003

A

A'Hearn, Michael F., 97, Fac, USA, ma@astro.umd.edu
Agrawal, Vivek K., 01, India, vivekag@isac.ernet.in
Ahn, Eun-Joo, 99, S. Korea, sein@oddjob.uchicago.edu
Ahumada, Andrea Veronica, 99, Argentina, andreahumada@hotmail.com
Akalin Peletier, Ayvur, 93, Turkey, ayvur_peletier@hotmail.com
Akujor, Chidi Emmanuel, 86, Nigeria, chiakujor@yahoo.com
Alonso Giraldes, Maria Sol, 01, Argentina, msolalonso@latinmail.com
Alonso-Herrero, Almudena, 93, Spain, aalonso@as.arizona.edu
Andreeva-Comerón, Petia, 95, Bulgaria, fcomeron@eso.org
Antipova, Jelena, 99, Estonia, jelena@physic.ut.ee
Aungwerojwit, Amornrat, 01, Thailand, amornrata@nu.ac.th

B

Baleisis, Audra, 93, USA, audra@as.arizona.edu
Bastian, Nate, 03, USA, nbastian@eso.org
Beaulieu, Sylvie, 90, Canada, sbeaulieu@phy.ulaval.ca
Beck, Judith A., 86, USA, jbeck@unca.edu
Beletsky, Yuri, 99 Ukraine, yuri@deneb.odessa.ua
Benaglia, Paula, 90, Argentina, paula@irma.iar.unlp.edu.ar
Berinde, Stefan, 97, Romania, sberinde@math.ubbcluj.ro
Biermann, Peter, 93, Fac, Germany, p165bie@mpifr-bonn.mpg.de
Biscaya, Ana Maria, 97, Portugal, anaholzbach@yahoo.com
Blokhina, Marina, 7, Ukraine, marina@astron.kharkov.ua
Bogosavljevic, Milan, 01, Serbia, milan@astro.caltech.edu
Bongiovanni, Angel M., 97, Venezuela, abongio@conicit.ve
Bornancini, Carlos G., 03, Argentina, bornancini@mail.oac.uncor.edu
Boyle, Richard, S.J., 93, 95, Fac, USA, boyle@as.arizona.edu
Briceno Avila, Cesar, 93, Venezuela, briceno@cida.ve
Brown, David A., S.J., 95, USA, davidbrownsj@jesuits.net
Bureau, Martin, 95, Canada, martin@mso.anu.edu.au

C

Callegari-Barlow, Rita, 86-95, Sec, Italy, barlowd@mediasoft.net
Campins, Humberto, 97, Fac, Venezuela, campins@physics.ucf.edu
Carreira, Emmanuel, S.J., 88-95, Fac, Spain, ecarreira@res.upco.es
Cassiano, Marta Maria, 88, Brazil, cassiano@if.usp.br
Castro, Sandra, 95 Brazil, scastro@astro1.iagusp.usp.br
Celletti, Alessandra, 86, Italy, celletti@axp.mat.uniroma2.it
Chakraborty, Pavan, 97, India, pavan@iiap.ernet.in
Chavez, Miguel, 90, Mexico, mchavez@inaoep.mx
Chen, An-Le, 86, Taiwan, alchen@scc.ntnu.edu.tw

Chen, Jun, 93, China
Cheng, Pei-Long, 88, Taiwan
Chernyshova, Irene V., 95, Ukraine, chernyshova@mail.ru
Chukwude, Augustine, 99, Nigeria, austine@bootes.hartrao.ac.za
Cimatti, Andrea, 90, Italy, cimatti@arcetri.astro.it
Comerón Tejero, Fernando, 90, Spain, fcomeron@eso.org
Consolmagno, Guy, S.J., 93-97, Fac, USA, gjc@as.arizona.edu
Cooper, Michael C., 01, USA, cooper@astron.berkeley.edu
Cora, Sofía, 93, Argentina, sacora@fcaglp.fcaglp.unlp.edu.ar
Corbally, Chris, S.J., 88-99, Fac, UK, corbally@as.arizona.edu
Credner, Till, 97, Germany, credner@unix1.mpae.gwdg.de
Cress, Catherine, 95, South Africa, cressc@nu.ac.za
Cuadra, Jorge, 01, Chile, jcuadra@mpa-garching.mpg.de
Curiel, Salvador R., 88, Mexico, scuriel@astroscu.unam.mx
Cvitan, Maro, 01, Croatia, mcvitan@diamond.phy.hr

D

D'Alessio, P., 88,Venezuela, p.dalessio@astrosmo.unam.mx
D'Antona, Francesca, 01, Fac, Italy, dantona@coma.mporzio.astro.it
Dapergolas, Anastasios, 86, Greece, adaperg@titan.astro.noa.gr
da Silva, Jose Ronaldo, 99, Brazil, ronaldo@dfte.ufrn.br
Dayanand, Sharmila, 97, S. Africa, sharmila@bootes.hartrao.ac.za
Devasagayam, M., S.J., 93, India, devasagayam@hotmail.com
Dey, Arjun, 86, India, dey@noao.edu
Dobrzycka, Danuta Panek, 90, Poland, danuta@cfa.harvard.edu
Dobrzycki, Adam, 86, Poland, adam@head-cfa.cfa.harvard.edu
Dominis, Dijana, 99, Croatia, ddominis@geodet.geof.hr
Donoso, Emilio, Argentina, edonoso@sinectis.com.ar
dos Anjos, Sandra, 86, Brazil, sandra@astro1.iagusp.usp.br
dos Santos, Igor Felipe, 03, Brazil, igor@dfte.ufrn.br
Dougherty, Sean M., 88, Canada, smd@drao.nrc.ca
Drouin, Nathalie, 93, Canada, drouin@astro.umontreal.ca
Durgapal, Alok Kumar, 95, India, akdurgapal@yahoo.co.in

E

Eenens, Philippe, 88, Belgium, eenens@avantel.net
Egami, Eiichi, 90, Japan, eegami@as.arizona.edu
Erspamer, Daniel, 99, Switzerland, daniel.erspamer@iastro.unil.ch
Espada, Daniel, 03, Spain, daniel@iaa.es
Evans, Aaron, 90, USA, aevans@vulcan.ess.sunysb.edu

F

Falcke, Heino, 93, Germany, hfalcke@mpifr-bonn.mpg.de
Feinstein, Carlos, 88, Argentina, cfeinstein@fcaglp.edu.ar
Folatelli, G., 99, Argentina, gaston@lahuan.fcaglp.unlp.edu.ar

Forbes, Duncan Alan, 86, New Zealand, dforbes@swin.edu.au
Foss, Diana, 86, USA, diana@mooseyard.com
Franco, Isabel, 03, Mexico, franco@fisica.unam.mx
Frenk, Carlos, 95, Fac, Argentina, c.s.frenk@durham.ac.uk
Funes, José, S.J., 93, Argentina, jfunes@as.arizona.edu
Fynbo, Johan, 95, Denmark, jfynbo@phys.au.dk

G

Galea, Cristina, 01, Romania, cgalea@mpifr-bonn.mpg.de
Galligan, David, 97, N. Zealand, d.galligan@phys.canterbury.ac.nz
Garaj, Slaven, 97, Croatia, slaven.garaj@epfl.ch
García, Javier A., 01, Venezuela, jgarcia@cida.ve
Garrison, Robert, 90, Fac, USA, garrison@astro.utoronto.ca
Gauba, Geetanjali, 99, India, gauba@iiap.ernet.in
Gazeas, Kosmas, 99, Greece, kgaze@skiathos.physics.auth.gr
Georganopoulos, Markos, 88, Greece, markos@mickey.mpi-hd.mpg.de
Gibson, Brad K., 88, Canada, bgibson@swin.edu.au
Gieles, Mark, 03, Netherlands, gieles@phys.uu.nl
Girish, T. E., 86, India, spkcv@md2.vsnl.net.in
Gomez, Mercedes, 88, Argentina, mergom@uncbob.edu.ar
Gorlova, Nadezhda, 99, Ukraine, nadir@deneb.odessa.ua
Gratsias, John, 88, Greece
Gray, Richard, 99, Fac, grayro@appstate.edu
Greve, Thomas Rodriguez, 99, Denmark, tgreve@astro.ku.dk
Guarnieri, Maria Donata, 86, Italy, mdg@star.roe.ac.uk
Gutierrez-Resendiz, Fidel, 01, Mexico, Fidel.Gutierrez-Resendiz@astro.cf.ac.uk

H

Hanninen, Jyrki, 88, Finland, jyrki@seita.oulu.fi
Hanski, Mikko, 95, Finland, mihanski@cc.utu.fi
Helmi, Amina, 95, Argentina, a.helmi@phys.uu.nl
Heynderickx, Daniel, 86, Belgium, daniel.heynderickx@ma.be
Hollywood, John B. S.J., 86, 88, USA, jbhgu@aol.com

I

Ilic, Dragana, 01, Serbia, ida7@EUnet.yu
Impey, Chris, 95, Fac, USA, cimpey@as.arizona.edu
Isik, Emre, 99, Turkey, eis@pascal.sci.akdeniz.edu.tr
Ivanov, Valentin, 93, Bulgaria, vivanov@eso.org
Ivanova, Albena, 97, Bulgaria, albena@optics.bas.bg

J

Jayawardhana, Ray, 93, Sri Lanka, rayjay@astro.lsa.umich.edu
Josephs, Tammy, 99, USA, tsimons.phys97@gtalumni.org

K

Kanaan Neto, Antonio, 88, Brazil, kanaan@if.ufrgs.br
Karciauskas, Mindaugas, 03, Lithuania, Mindaugas.Karciauskas@ff.vu.lt
Kaufman Bernardó, Maria Marina, 01, Argentina, mar@vianw.net.ar
Keppel, Jean W., 88, Fac, USA
Keskin, Varol, 88, Turkey, keskinv@alpha.sci.ege.edu.tr
Khoi, Dinh Phan, 95, Vietnam, khoi@mesioq.obspm.fr
Kilinc, B. 90, Turkey, efeast01%trearn.bitnet@cunyvm.cuny.edu
Kim, Tae-Sun, 90, Korea, tkim@eso.org
Kim, Yongha, 86, Korea
Kimeswenger, Stefan, 86, Austria, stefan.kimeswenger@uibk.ac.at
Kjoor, Karina, 01, Denmark, karina@astro.ku.dk
Kohnenkamp, Ive A., 95, Chile, ikohnenkamp@as.arizona.edu
Kolhatkar, Sonali, 95, India
Koposov, Sergey, 03, Russia, math@sai.msu.ru
Krasteva, Eleonora, 95, Bulgaria, eleonora@umich.edu
Kravtsov, Andrey V., 95, Russia, andrey@astronomy.ohio-state.edu
Kube, Jens, 99, Germany, jkube@awi-potsdam.de
Kuipers, Dorothy, 03, UK, d.kuipers@ucl.ac.uk

L

Lacerda, Reiner Rodrigues, 99, Brazil, reiner@fsc.ufsc.br
Lada, Charles, 88, Fac, USA, clada@cfa.harvard.edu
Latham, David, 86, Fac, USA, latham@cfa.harvard.edu
Lederer, Susan M., 97, USA, slederer@csusb.edu
Leka, Kimberly Dawn, 88, USA, leka@colorado-research.com
Li, Aigen, 97, Rep. of China, agli@lpl.arizona.edu
Li, Moping, 03, China, limoping@sina.com
López-Cruz, Omar, 88, Mexico, omar@inaoep.mx
Luhman, Kevin L., 93, USA, kluhman@cfa.harvard.edu
Lyubenova, Mariya, 03, Bulgaria, m_lyubenova@yahoo.com

M

Ma, Bo-Qum, 90, Rep. of China
Maciejewski, Witold, 93, Poland, witold@oa.uj.edu.pl
MacPherson, Alick Lachlan, 88, Canada, alickm@mail.cern.ch
Maggio, Elizabeth, 97, Sec, USA, lizmaggio@compuserve.com
Magris Crestini, Gladis, 88, Venezuela, gmagris@cida.ve
Maiolino, Roberto, 93, Italy, maiolino@arcetri.astro.it
Mardones, Diego, 90, Chile, diego@das.uchile.cl

Marin, Zaida, 90, Venezuela, marin@cida.ve
Mashchenko, Sergey Y., 93, Ukraine, syam@phy.ulaval.ca
Matsoukas, Christos, 93, Greece, cmatsouk@ddt.eng.uc.edu
McCarthy, Martin F., 86, Fac, USA
Mechtcheriakov, Alexandre, 01, Russia, shura@xray.sai.msu.ru
Medina Tanco, G., 90, Argentina, gustavo@astro1.iagusp.usp.br
Medupe, Thebe, 95, S. Africa, thebe@saao.ac.za
Melbourne, Jason, 01, USA, jmelbourne@wesleyan.edu
Melkonian, Anahit, 95, Armenia, anahit@lx2.yerphi.am
Melo, Claudio de Figueiredo, Brazil, hclaudio@dfte.ufrn.br
Miah, Muhammad Adel, 86, Iran
Milin, Matko, 95, Croatia, milin@olimp.irb.hr
Milone, Alejandra, 86, Argentina, amilone@as.arizona.edu
Minniti, Dante, 86, Argentina, dante@astro.puc.cl
Mokler, Felicitas, Germany, fmokler@mpe.mpg.de
Molenda-Zakowicz, Joanna, 99, Poland, molenda@ifa.au.dk
Molina, Faviola, 03, Venezuela, fmolina@cida.ve
Momcheva, Ivelina G., 01, Bulgaria, imomcheva@as.arizona.edu
Moreno Méndez, Enrique, 01, Mexico, enriquemm73@yahoo.com
Mossman, Amy, 90, USA, amossman@head-cfa.harvard.edu
Motta, Veronica, 97, Uruguay, motta@fisica.edu.uy
Mújica, Raul, 93, Mexico, rmujica@inaoep.mx
Mullis, Christopher, 95, USA, mullis@galileo.ifa.hawaii.edu
Munasinghe, Rohan, 97, Sri Lanka, rohan@odyssey.kaist.ac.kr
Murdoch, Kaylene, 90, New Zealand, kaylene_murdoch@stats.govt.nz
Mysore, Sangeeta, 90, India

N

Najita, Joan, 88, USA, najita@noao.edu
Narayan, Ramesh, 01, Fac, India, narayan@cfa.harvard.edu

O

Okorogu, Albert O.C., 90, Nigeria, albert.okorogu@amadon.com
Okpala, Kingsley C., 97, Nigeria, okpala@infoweb.abs.net
Orellana, Mariana, 03, Argentina, morellana@carina.fcaglp.unlp.edu.ar

P

Pakzad, Sabrina, 03, USA, sabpakzad@yahoo.com
Palma, Norman I., 01, Honduras, npalma17@hotmail.com
Pavlov, Alexander A., 97, Russia, pavlov@lasp.colorado.edu
Perera, Lalith, 90, Sri Lanka, perera@lns62.lns.cornell.edu
Pereyra, Antonio, 93, Peru, antonio@argus.iagusp.usp.br
Pesce-Rollins, Melissa, 97, Sec., Italy, campiglia44@hotmail.com
Petkaki, Panagiota G., 90, Greece, PPE@bas.ac.uk
Petrovay, Kristof, 88, Hungary, k.petrovay@astro.elte.hu

Piirola, Vilppu, 88, 90, Fac, Finland, piirola@sara.cc.utu.fi
Pilman, Christopher, 01, USA, cpilman@cfa.harvard.edu
Pipino, Antonio, 03, Italy, antonio@ts.astro.it
Poirier, Sara, 97, Canada, sara.poirier@osc.on.ca
Pollard, Karen, 90, N. Zealand, k.pollard@phys.canterbury.ac.nz
Puss, Alar, 99, Finland, alar@aai.ee

Q

Qiu, Yulei, 99, China, qiuyl@nova.bao.ac.cn

R

Rao, Meenakshi, 86, India, meenakshi_rao@yahoo.com
Ratag, Mezak Arnold, 88, Indonesia, lapan@hotmail.com
Reese, Vladimir A. Avila, 95, Colombia, avila@astroscu.unam.mx
Rejkuba, Marina, 97, Croatia, mrejkuba@eso.org
Richards, Mercedes, 99, Fac, richards@astro.psu.edu
Rieke, George, 93, Fac, USA, grieke@as.arizona.edu
Rieke, Marcia, 93, Fac, USA, mrieke@as.arizona.edu
Rodrigues, Claudia Vilega, 93, Brazil, claudia@das.inpe.br
Rodriguez Flores, Ernesto R., 99, Cuba, iga@cidet.icmf.inf.cu
Romano, Patrizia, 95, Italy, romano@merate.mi.astro.it
Romero, Gustavo E., 93, Argentina, romero@irma.iar.unlp.edu.ar
Rosales, Fernando Fabián, 03, Mexico, fer_fab@hotmail.com
Rubin, Vera C., 86, Fac, USA, rubin@gal.ciw.edu
Ruiz, Milagros, 90, Peru, mili@star.herts.ac.uk

S

Saint-Jacques, David, 99, Canada, dastj14@agora.ulaval.ca
Sägesser, Sarah N., 01, Switzerland, sagesser@thphys.ox.ac.uk
Sandin, Christer, 97, Sweden, xrisp@grub01.physto.se
Sankarankutty, Shobha, 97, Brazil, shobha.sankarankutty@anu.edu.au
Sarajedini, Ata, 86, USA, ta@astro.ufl.edu
Sargsysan, Lousine A., 01, Armenia, Sarl11@yahoo.com
Scafuro, Giovanni, 86, Italy
Schlaerth, James, 03, USA, james.a.schlaerth@vanderbilt.edu
Schmidt, Gary D., 01, Fac, USA, gschmidt@as.arizona.edu
Seeberger, Robert, 90, Austria
Serenelli, Aldo, 01, Argentina, serenell@fcaglp.fcaglp.unlp.edu.ar
Seshabela, Rebecca, 01, South Africa, mseshabela@yahoo.com
Shu, Frank, 88, Fac, USA, shu@yastro.berkeley.edu
Smolcic, Vernesa, 03, Croatia, vernesa@norma.hr
Soto, Alejandro, 97, USA, Alejandro.M.Soto.97@Alum.Dartmouth.org
Steren, Gabriela A., 97, Argentina, gabriela.uncbob.edu.ar
Stocke, John T., 90, Fac, USA, stocke@casa.colorado.edu
Stoeger, William R., S.J., 95, Fac, USA, wstoeger@as.arizona.edu

Strateva, Iskra, 99, Bulgaria, iskra@astro.princeton.edu
Streblyanskaya, Alina, 03, Ukraine, alina@mpe.mpg.de

T

Taylor, Andrew David, 86, New Zealand
Timokhin, Andrew N., 95, Russia, Andrey.Timokhin@mpi-hd.mpg.de
Torres, Diego, 95, Argentina, dtorres@igpp.ucllnl.org
Tóth, Juraj, 97, Slovakia, toth@center.fmph.uniba.sk
Trammell, Susan Rebecca, 88, USA, srtramme@uncc.edu
Tripe, Peter, 93, South Africa, trppet01@uctvax.uct.ac.za

V

Valiauga, Gintaras, 93, Lithuania, valiauga.astro@itpa.fi.lt
Van Gorkom, Jacqueline, 90, Fac, USA, jvangork@astro.columbia.edu
Vanhala, Harri, 90, Finland, harri@hiisi.oulu.fi
Vanzi, Leonardo, 93, Italy, lvanzi@eso.org
van Zyl, Liza, 97, South Africa, lvz@astro.keele.ac.uk
Vasile, Ana, 03, Romania, avasile@mpifr-bonn.mpg.de
Verley, Simon, 03, France, simon@iaa.es
Verronen, Maarit Hannele, 88, Finland, maarit@hiisi.oulu.fi
Vidrih, Simon, 01, Slovenia, simon.vidrih@fmf.uni-lj.si
Vieira, Katherine, 01, Venezuela, vieira@astro.yale.edu, kvieira@cida.ve
Vilas, Faith, 97, Fac, USA, faith.vilas1@jsc.nasa.gov
Villegas, Daniela Muri, 03, Chile, dmvilleg@puc.cl
Vivas, Anna K., 95, Venezuela, vivas@cida.ve

W

Wang, Yiping, 93, China, ypwang@pmo.ac.cn
Wauben, Wiel, 86, The Netherlands
West, Steve R. D., 93, New Zealand
Wickramasinghe, Thulsi, 88, Sri Lanka, wick@astro.tcnj.edu
Wiegert, Theresa, 03, Sweden, theresaw@oso.chalmers.se
Wilner, David James, 88, USA, wilner@cfa.harvard.edu
Wuyts, Stijn, 01, Belgium, wuyts@phys.leidenuniv.nl
Wylie, Elizabeth, 03, New Zealand, ECW21@student.canterbury.ac.nz

X

Xiluri, Kiriaki, 86, Greece, kx8u@carina.astro.virginia.edu

Z

Zamojski, Michel 03, Canada, michel@astro.columbia.edu
Zembrowski, Piotr, 90, Poland
Zwitter, Tomaz, 86, Slovenia, tomaz.zwitter@uni-lj.si

Author Index

Subject Index

Colophon

The articles in Part 1 of these Proceedings were prepared by the contributing authors, who primarily used Microsoft Word. The formatting of Part 1 was performed by the editors, who exclusively used Microsoft Word. The articles in Part 2 of these Proceedings were prepared by the contributing authors using LaTeX in the style of the Publications of the Astronomical Society of the Pacific, `paspconf`. Assembly of this book (except for Part 1) was performed using editorial software tools, `pdb`, written by Chris Biemesderfer (ferberts associates) and made available by Jeannett Barnes (NOAO). These scripts process the individual articles and compile the author names and titles into the table of contents, the conference participants list, and author and subject indices referencing the appropriate page numbers. Place holders were used to map the articles from Part 1 into the front and back matter.

VATICAN PRESS